Cambridge Elements ≡

Elements in the Structure and Dynamics of Complex Networks
edited by
Guido Caldarelli
Ca' Foscari University of Venice

GILLESPIE ALGORITHMS FOR STOCHASTIC MULTIAGENT DYNAMICS IN POPULATIONS AND NETWORKS

Naoki Masuda
*State University of New York at Buffalo
and Waseda University*

Christian L. Vestergaard
Institut Pasteur, CNRS, Paris

CAMBRIDGE
UNIVERSITY PRESS

CAMBRIDGE
UNIVERSITY PRESS

Shaftesbury Road, Cambridge CB2 8EA, United Kingdom

One Liberty Plaza, 20th Floor, New York, NY 10006, USA

477 Williamstown Road, Port Melbourne, VIC 3207, Australia

314–321, 3rd Floor, Plot 3, Splendor Forum, Jasola District Centre,
New Delhi – 110025, India

103 Penang Road, #05–06/07, Visioncrest Commercial, Singapore 238467

Cambridge University Press is part of Cambridge University Press & Assessment,
a department of the University of Cambridge.

We share the University's mission to contribute to society through the pursuit of
education, learning and research at the highest international levels of excellence.

www.cambridge.org
Information on this title: www.cambridge.org/9781009239141
DOI: 10.1017/9781009239158

First published 2022

A catalogue record for this publication is available from the British Library.

ISBN 978-1-009-23914-1 Paperback
ISSN 2516-5763 (online)
ISSN 2516-5755 (print)

Gillespie Algorithms for Stochastic Multiagent Dynamics in Populations and Networks

Elements in the Structure and Dynamics of Complex Networks

DOI: 10.1017/9781009239158
First published online: December 2022

Naoki Masuda
State University of New York at Buffalo and Waseda University

Christian L. Vestergaard
Institut Pasteur, CNRS, Paris

Author for correspondence: Naoki Masuda, naokimas@buffalo.edu and Christian Vestergaard, christian.vestergaard@cnrs.fr

Abstract: Many multiagent dynamics can be modeled as a stochastic process in which the agents in the system change their state over time in interaction with each other. The Gillespie algorithms are popular algorithms that exactly simulate such stochastic multiagent dynamics when each state change is driven by a discrete event, the dynamics are defined in continuous time, and the stochastic law of event occurrence is governed by independent Poisson processes. The first main part of this Element provides a tutorial on the Gillespie algorithms focusing on simulation of social multiagent dynamics occurring in populations and networks. The authors clarify why one should use the continuous-time models and the Gillespie algorithms in many cases instead of easier-to-understand discrete-time models. The remainder of the Element reviews recent extensions of the Gillespie algorithms aiming to add more reality to the model (i.e., non-Poissonian cases) or to speed up the simulations. This title is also available as open access on Cambridge Core.

Keywords: numerical simulations, jump processes, Poisson processes, renewal processes, complex systems

ISBNs: 9781009239141 (PB), 9781009239158 (OC)
ISSNs: 2516-5763 (online), 2516-5755 (print)

Contents

1 Introduction

We are compelled to understand and intervene in the dynamics of various complex systems in which different elements, such as human individuals, interact with each other. Such complex systems are often modeled by multiagent or network-based models that explicitly dictate how each individual behaves and influences other individuals. Stochastic processes are popular models for the dynamics of multiagent systems when it is realistic to assume random elements in how agents behave or in dynamical processes taking place in the system. For example, random walks have been successfully applied to describe locomotion and foraging of animals (Codling, Plank, & Benhamou, 2008; Okubo & Levin, 2001), dynamics of neuronal firing (Gabbiani & Cox, 2010; Tuckwell, 1988), and financial market dynamics (Campbell, Lo, & MacKinlay, 1997; Mantegna & Stanley, 2000) to name a few (see Masuda, Porter, and Lambiotte [2017] for a review). Branching processes are another major type of stochastic processes that have been applied to describe, for example, information spread (Eugster et al., 2004; Gleeson et al., 2021), spread of infectious diseases (Britton, 2010; Farrington, Kanaan, & Gay, 2003), cell proliferation (Jagers, 1975), and the abundance of species in a community (McGill et al., 2007) as well as other ecological dynamics (Black & McKane, 2012).

Stochastic processes in which the state of the system changes via discrete events that occur at given points in time are a major class of models for dynamics of complex systems (Andersson & Britton, 2000; Barrat, Barthélemy, & Vespignani, 2008; Daley & Gani, 1999; de Arruda, Rodrigues, & Moreno, 2018; Kiss, Miller, & Simon, 2017a; Liggett, 2010; Shelton & Ciardo, 2014; Singer & Spilerman, 1976; Van Mieghem, 2014). For example, in typical models for infectious disease spread, each infection event occurs at a given time t such that an individual transitions instantaneously from a healthy to an infectious state. Such processes are called *Markov jump processes* when they satisfy certain independence conditions (Hanson, 2007), which we will briefly discuss in Section 2.5. A jump is equivalent to a discrete event. In Markov jump processes, jumps occur according to *Poisson processes*. In this volume, we focus on how to simulate Markov jump processes. Specifically, we will introduce a set of exact and computationally efficient simulation algorithms collectively known as Gillespie algorithms. In the last technical section of this volume (i.e., Section 5), we will also consider more general, non-Markov, jump processes, in which the events are generated in more complicated manners than by Poisson processes. In the following text, we refer collectively to Markov jump processes and non-Markov jump processes as jump processes.

The Gillespie algorithms were originally proposed in their general forms by Daniel Gillespie in 1976 for simulating systems of chemical reactions (Gillespie, 1976), whereas several specialized variants had been proposed earlier; see Section 3.1 for a brief history review. Gillespie proposed two different variants of the simulation algorithm, the *direct method*, also known as Gillespie's *stochastic simulation algorithm* (SSA), or often simply *the Gillespie algorithm*, and the *first reaction method*. Both the direct and first reaction methods have found widespread use and in fields far beyond chemical physics. Furthermore, researchers have developed many extensions and improvements of the original Gillespie algorithms to widen the types of processes that we can simulate with them and to improve their computational efficiency.

The Gillespie algorithms are practical algorithms to simulate coupled Poisson processes exactly (i.e., without approximation error). Here "coupled" means that an event that occurs somewhere in the system potentially influences the likelihood of future events' occurrences in different parts of the same system. For example, when an individual in a population, v_i, gets infected by a contagious disease, the likelihood that a different healthy individual in the same population, v_j, will get infected in the near future may increase. If interactions were absent, it would suffice to separately consider single Poisson processes, and simulating the system would be straightforward.

We believe that the Gillespie algorithms are important tools for students and researchers that study dynamic social systems, where social dynamics are broadly construed and include both human and animal interactions, ecological systems, and even technological systems. While there already exists a large body of references on the Gillespie algorithms and their variants, most are concise, mathematically challenging for beginners, and focused on chemical reaction systems.

Given these considerations, the primary aim of this volume is to provide a detailed tutorial on the Gillespie algorithms, with specific focus on simulating dynamic social systems. We will realize the tutorial in the first part of the Element (Sections 2 and 3). In this part, we assume basic knowledge of calculus and probability. Although we do introduce stochastic processes and explain the Gillespie algorithms and related concepts with much reference to networks, we do not assume prior knowledge of stochastic processes or of networks. To understand the coding section, readers will need basic knowledge of programming. The second part of this Element (Sections 4 and 5) is devoted to a survey of recent advancements of Gillespie algorithms for simulating social dynamics. These advancements are concerned with accelerating simulations and/or increasing the realism of the models to be simulated.

2 Preliminaries

We review in this section mathematical concepts needed to understand the Gillespie algorithms. In Sections 2.1 to 2.3, we introduce the types of models we will be concerned with, namely *jump processes*, and in particular a simple type of jump process termed *Poisson processes*. In Sections 2.4 to 2.6, we derive the main mathematical properties of Poisson processes. The concepts and results presented in Sections 2.1 to 2.6 are necessary for understanding Section 3, where we derive the Gillespie algorithms. In Sections 2.7 and 2.8, we review two simple methods for solving the models that predate the Gillespie algorithms and discuss some of their shortcomings. These two final subsections motivate the need for exact simulation algorithms such as the Gillespie algorithms.

2.1 Jump Processes

Before getting into the nitty-gritty of the Gillespie algorithms, we first explore which types of systems they can be used to simulate. First of all, with the Gillespie algorithms, we are interested in simulating a dynamic system. This can be, for example, epidemic dynamics in a population in which the number of infectious individuals varies over time, or the evolution of the number of crimes in a city, which also varies over time in general. Second, the Gillespie algorithms rely on a predefined and parametrized mathematical model for the system to simulate. Therefore, we must have the set of rules for how the system or the individuals in it change their states. Third, Gillespie algorithms simulate stochastic processes, not deterministic systems. In other words, every time one runs the same model starting from the same initial conditions, the results will generally differ. In contrast, in a deterministic dynamical system, if we specify the model and the initial conditions, the behavior of the model will always be the same. Fourth and last, the Gillespie algorithms simulate processes in which changes in the system are primarily driven by discrete events taking place in continuous time. For example, when a chemical reaction obeying the chemical equation $A + B \rightarrow C + D$ happens, one unit each of A and of B are consumed, and one unit each of C and of D are produced. This event is discrete in that we can count the event and say when the event happened, but it can happen at any point in time (i.e., time is not discretized but continuous).

We refer to the class of mathematical models that satisfy these conditions and may be simulated by a Gillespie algorithm as *jump processes*. In the remainder of this section, we explore these processes more extensively through motivating examples. Then, we introduce some fundamental mathematical definitions and results that the Gillespie algorithms rely on.

2.2 Representing a Population as a Network

Networks are an extensively used abstraction for representing a structured population, and Gillespie algorithms lend themselves naturally to simulate stochastic dynamical processes taking place in networks. In a network representation, each individual in the population corresponds to a node in the network, and edges are drawn between pairs of individuals that directly interact. What constitutes an interaction generally depends on the context. In particular, for the simulation of dynamic processes in the population, the interaction depends on the nature of the process we wish to simulate. For simulating the spread of an infectious disease, for example, a typical type of relevant interaction is physical proximity between individuals.

Formally, we define a network as a graph $G = (V, E)$, where $V = \{1, 2, \ldots, N\}$ is the set of nodes, $E = \{(u, v): u, v \in V\}$ is the set of edges, and each edge (u, v) defines a pair of nodes $u, v \in V$ that are directly connected. The pairs (u, v) may be ordered, in which case edges are directed (by convention from u to v), or unordered, in which case edges are undirected (i.e., v connects to u if and only if u connects to v). We may also add weights to the edges to represent different strengths of interactions, or we may even consider graphs that evolve in time (so-called temporal networks) to account for the dynamics of interactions in a population.

We will primarily consider simple (i.e., static, undirected, and unweighted) networks in our examples. However, the Gillespie algorithms apply to simulated jump processes in all kinds of populations and networks. (For temporal networks, we need to extend the classic Gillespie algorithms to cope with the time-varying network structure; see Section 5.4.)

2.3 Example: Stochastic SIR Model in Continuous Time

We introduce jump processes and explore their mathematical properties by way of a running example. We show how we can use them to model epidemic dynamics using the stochastic susceptible-infectious-recovered (SIR) model.[1] For more examples (namely, SIR epidemic dynamics in metapopulation networks, the voter model, and the Lotka–Volterra model for predator–prey dynamics), see Section 3.4.

We examine a stochastic version of the SIR model in continuous time defined as follows. We consider a constant population of N individuals (nodes). At any time, each individual is in one of three states: susceptible (denoted by S;

---

[1] The SIR model was incidentally one of the first applications of a Gillespie-type algorithm in a 1953 article (Bartlett, 1953).

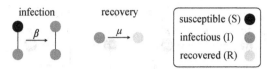

Figure 1 Rules of state changes in the SIR model. An infectious individual infects a susceptible neighbor at a rate β. Each infectious individual recovers at a rate μ.

meaning healthy), infectious (denoted by I), or recovered (denoted by R). The rules governing how individuals change their states are shown schematically in Fig. 1. An infectious individual that is in contact with a susceptible individual infects the susceptible individual in a stochastic manner with a constant *infection rate β*. Independently of the infection events, an infectious individual may recover at any point in time, with a constant *recovery rate μ*. If an infection event occurs, the susceptible individual that has been infected changes its state to I. If an infectious individual recovers, it transits from the I state to the R state. Nobody leaves or joins the population over the course of the dynamics. After reaching the R state, an individual cannot be reinfected or infect others again. Therefore, R individuals do not influence the increase or decrease in the number of S or I individuals. Because R individuals are as if they no longer exist in the system, the R state is mathematically equivalent to having died of the infection; once dead, an individual will not be reinfected or infect others.

We typically start the stochastic SIR dynamics with a single infectious individual, which we refer to as the *source* or *seed*, and $N_S = N - 1$ susceptible individuals (and thus no recovered individuals). Then, various infection and recovery events may occur. The dynamics stop when no infectious individuals are left. In this final situation, the population is composed entirely of susceptible and/or recovered individuals. Since both infection and recovery involve an infectious individual, and there are no infectious individuals left, the dynamics are stuck. The final number of recovered nodes, denoted by N_R, is called the *epidemic size*, also known as the *final epidemic size* or simply the *final size*.[2] The epidemic size tends to increase as the infection rate β increases or as the recovery rate μ decreases. Many other measures to quantify the behavior of the SIR model exist (Pastor-Satorras et al., 2015). For example, we may be interested in the time until the dynamics terminate or in the speed at which the number of infectious individuals grows in the initial stage of the dynamics.

---

[2] The fraction N_R/N is typically also referred to as the epidemic size.

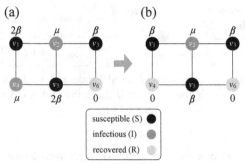

Figure 2 Stochastic SIR process on a square-grid network with six nodes. (a) Status of the network at an arbitrary time *t*. (b) Status of the network after v_4 has recovered. The values attached to the nodes indicate the rates of the events that the nodes may experience next.

Consider Fig. 2(a), where individuals are connected as a network. We generally assume that infection may only occur between pairs of individuals that are directly connected by an edge (called adjacent nodes). For example, the node v_4 can infect v_1 and v_5 but not v_3. The network version of the SIR model is fully described by the infection rate β, the recovery rate μ, the network structure, that is, which node pairs are connected by an edge, and the choice of source node to initialize the dynamics.

Mathematically, we describe the system by a set of coupled, constant-rate jump processes; constant-rate jump processes are known as *Poisson processes* (Box 1). Each possible event that may happen is associated to a Poisson process, that is, the recovery of each infectious individual is described by a Poisson process, and so is each pair of infectious and susceptible individuals where the former may infect the latter. The Poisson processes are coupled because an event generated by one process may alter the other processes by changing their rates, generating new Poisson processes, or making existing ones disappear. For example, after a node gets infected, it may in turn infect any of its susceptible neighbors, which we represent mathematically by adding new Poisson processes. This coupling implies that the set of coupled Poisson processes generally constitutes a process that is more complicated than a single Poisson process.

In the following subsections we develop the main mathematical properties of Poisson processes and of sets of Poisson processes. We will rely on these properties in Section 3 to construct the Gillespie algorithms that can simulate systems of coupled Poisson processes exactly. Note that the restriction to Poisson (i.e., constant-rate) processes is essential for the classic Gillespie algorithms to work; see Section 5 for recent extensions to the simulation of non-Poissonian processes.

Box 1 Properties of Poisson Processes

A Poisson process is a jump process that generates events with a constant rate, λ.

Waiting-Time Distribution

The waiting times τ between consecutive events generated by a Poisson process are exponentially distributed. In other words, τ obeys the probability density

$$\psi(\tau) = \lambda e^{-\lambda \tau}. \tag{2.1}$$

Memoryless Property

The waiting time left until a Poisson process generates an event given that a time t has already elapsed since the last event is independent of t. This property is called the memoryless property of Poisson processes and is shown as follows:

$$\psi(t+\tau|t) = \frac{\psi(t+\tau)}{\Psi(t)} = \frac{\lambda e^{-\lambda(t+\tau)}}{e^{-\lambda t}} = \lambda e^{-\lambda \tau}, \tag{2.2}$$

where $\psi(t+\tau|t)$ represents the conditional probability density that the next event occurs a time $t + \tau$ after the last event given that time t has already elapsed; $\Psi(t) = \int_t^\infty \psi(\tau)d\tau = e^{-\lambda t}$ is called the survival probability and is the probability that no event takes place for a time t. The first equality in Eq. (2.2) follows from the definition of the conditional probability. The second equality follows from Eq. (2.1).

Superposition Theorem

Consider a set of Poisson processes indexed by $i \in \{1, 2, \ldots, M\}$. The superposition of the processes is a jump process that generates an event whenever any of the individual processes does. It is another Poisson process whose rate is given by

$$\Lambda = \sum_{i=1}^M \lambda_i, \tag{2.3}$$

where λ_i is the rate of the ith Poisson process.

Box 1 (Continued)

Probability of a Given Process Generating an Event in a Superposition of Poisson Processes

Consider any given event generated by a superposition of Poisson processes. The probability Π_i that the ith individual Poisson process has generated this event is proportional to the rate of the ith process. In other words,

$$\Pi_i = \lambda_i / \Lambda. \tag{2.4}$$

2.4 Waiting-Time Distribution for a Poisson Process

We derive in this subsection the *waiting-time distribution* for a Poisson process, which characterizes how long one has to wait for the process to generate an event. It is often easiest to start from a discrete-time description when exploring properties of a continuous-time stochastic process. Therefore, we will follow this approach here. We use the recovery of a single node in the SIR model as an example in our development.

Let us partition time into short intervals of length δt. As δt goes to zero, this becomes an exact description of the continuous-time process. An infectious individual recovers with probability $\mu \delta t$ after each interval given that it has not recovered before.[3]

Formally, we define the SIR process in the limit $\delta t \to 0$. Then, you might worry that the recovery event is unlikely to ever take place because the probability with which it happens during each time-step, that is, $\mu \delta t$, goes toward 0 when the step size δt does so. However, this is not the case; because the number of time-steps in any given finite interval grows inversely proportional to δt, the probability to recover in finite time stays finite. For example, if we use a different step size $\overline{\delta t} = \delta t / 10$, which is ten times smaller than the original δt, then the probability of recovery within the short duration of time $\overline{\delta t}$ is indeed 10 times smaller than $\mu \delta t$ (i.e., $= \mu \overline{\delta t}$). However, there are $\delta t / \overline{\delta t} = 10$ windows of size $\overline{\delta t}$ in one time window of size δt. So, we now have 10 chances for recovery to happen instead of only one chance. The probability for recovery to occur in any of these 10 time windows is equal to one minus the probability that it does not occur. The probability that the individual does not recover in time δt is equal

---

[3] To address a common misunderstanding, we emphasize that μ is a rate, not a probability, and thus can be larger than one. Note, however, that $\mu \delta t$ is a probability and thus cannot be greater than one.

to $(1 - \mu\overline{\delta t})^{\delta t/\overline{\delta t}}$. Therefore, the probability that the individual recovers in any of the $\delta t/\overline{\delta t}$ windows is

$$p_{I \to R} = 1 - (1 - \mu\overline{\delta t})^{\delta t/\overline{\delta t}}. \tag{2.5}$$

Equation (2.5) does not vanish as we make $\overline{\delta t}$ small. In fact, the Taylor expansion of Eq. (2.5) in terms of $\overline{\delta t}$ yields $p_{I \to R} \approx (\delta t/\overline{\delta t}) \times \mu\overline{\delta t} = \mu\delta t$, where $\approx$ represents "approximately equal to". Therefore, to leading order, the recovery probabilities are the same between the case of a single time window of size δt and the case of $\delta t/\overline{\delta t}$ time windows of size $\overline{\delta t}$.

In the limit $\delta t \to 0$, the recovery event may happen at any continuous point in time. We denote by τ the waiting time from the present time until the time of the recovery event. We want to determine the probability density function (probability density or pdf for short) of τ, which we denote by $\psi_{I \to R}(\tau)$. By definition, $\psi_{I \to R}(\tau)\delta t$ is equal to the probability that the recovery event happens in the interval $[\tau, \tau + \delta t)$ for an infinitesimal δt (i.e., for $\delta t \to 0$). To calculate $\psi_{I \to R}(\tau)$, we note that the probability that the event occurs after $r = \tau/\delta t$ time windows, denoted by $p_{I \to R}(r)$, is equal to the probability that it did not occur during the first r time windows and then occurs in the $(r + 1)$th window. This probability is equal to

$$p_{I \to R}(r) = (1 - \mu\delta t)^r \times \mu\delta t = (1 - \mu\delta t)^{\tau/\delta t}\mu\delta t. \tag{2.6}$$

The first factor on the right-hand side of Eq. (2.6) is the probability that the event has not happened before the $(r + 1)$th window; it is simply equal to the probability that the event has not happened during a single window, raised to the power of r. The second factor is the probability that the event happens in the $(r + 1)$th window. By applying the identity $\lim_{x \to 0}(1 + x)^{1/x} = e$, known from calculus (see Appendix), with $x = -\mu\delta t$ to Eq. (2.6), we obtain the pdf of the waiting time as follows:

$$\begin{aligned} \psi_{I \to R}(\tau) &= \lim_{\delta t \to 0} \frac{p_{I \to R}(\tau/\delta t)}{\delta t} \\ &= \mu \lim_{\delta t \to 0} (1 - \mu\delta t)^{\tau/\delta t} \\ &= \mu \left[\lim_{\delta t \to 0} (1 - \mu\delta t)^{1/(-\mu\delta t)} \right]^{-\mu\tau} \\ &= \mu e^{-\mu\tau}. \end{aligned} \tag{2.7}$$

Equation (2.7) shows the intricate connection between the Poisson process and the exponential distribution: the waiting time of a Poisson process with rate μ (here, specifically the recovery rate) follows an exponential distribution with rate μ (Box 1). This fact implies that the mean time we have to wait for the recovery event to happen is $1/\mu$. The exponential waiting-time distribution actually completely characterizes the Poisson process. In other words, the Poisson process is the only jump process that generates events separated by waiting times that follow a fixed exponential distribution.

If we consider the infection process between a pair of S and I nodes in complete isolation from the other infection and recovery processes in the population, then exactly the same argument (Eq. (2.7)) holds true. In other words, the time until infection takes place between the two nodes is exponentially distributed with rate β, that is,

$$\psi_{S \to I}(\tau) = \beta e^{-\beta \tau}. \tag{2.8}$$

However, in practice the infection process is more complicated than the recovery process because it is coupled to other processes. Specifically, if another process generates an event before the infection process does, then Eq. (2.8) may no longer hold true for the infection process in question. For example, consider a node v_1 that is currently susceptible and an adjacent node v_2 that is infectious, as in Fig. 2. For this pair of nodes, two events are possible: v_2 may infect v_1, or v_2 may recover. As long as neither of the events has yet taken place, either of the two corresponding Poisson processes may generate an event at any point in time, following Eqs. (2.8) and (2.7), respectively. However, if v_2 recovers before it infects v_1, then the infection event is no longer possible, and so Eq. (2.8) no longer holds. We explore in the following two subsections how to mathematically deal with this coupling.

2.5 Independence and Interdependence of Jump Processes

Most models based on jump processes and most simulation methods, including the Gillespie algorithms, implicitly assume that different concurrent jump processes are independent of each other in the sense that the internal state of one process does not influence another. This notion of independence may be a source of confusion because a given process may depend on the events generated earlier by other processes, that is, the processes may be *coupled*, as we saw is the case for the infection processes in the SIR model. In this section, we sort out the notions of independence and coupling and what they mean for the types of jump processes we want to simulate. We will also explore another type of independence of Poisson processes, which is their independence of the past, called the *memoryless* property.

We can state the independence assumption as the condition that different processes are only allowed to influence each other by changing the state of the system. In other words, at any point in time each process generates an event at a rate that is independent of all other processes given the current state of the system, that is, the processes are *conditionally independent*. For example, the rate at which v_2 infects v_1 in Fig. 2(a) depends on v_2 being infectious and v_1 being susceptible (corresponding to the system's current state). However, it does not depend on any internal state of v_2's recovery process such as the time left until v_2 recovers. Given the states of all nodes, the two processes are independent. Poisson processes are always conditionally independent in this sense. The conditional independence property follows directly from the fact that Poisson processes have constant rates by definition and thus are not influenced by other processes. The conditional independence is essential for the Gillespie algorithms to work. Even the recent extensions of the Gillespie algorithms to simulate non-Poissonian processes, which we review in Section 5, rely on an assumption of conditional independence between the jump processes.

We underline that the assumption of conditional independence does not imply that the different jump processes are not coupled with each other. Such uncoupled processes would indeed be boring. If the jump processes constituting a given system were all uncoupled, then they would not be able to generate any collective dynamics. On the technical side, there would in this case be no reason to consider the set of processes as one system. It would suffice to analyze each process separately. We would in particular have no need for the specialized machinery of the Gillespie algorithms since we could simply simulate each process by sampling waiting times from the corresponding exponential distribution (Box 1, Eq. (2.1)).

In fact, the conditional independence assumption allows different processes to be coupled, as long as they only do so by changing the physical state of the system. This is a natural constraint in many systems. For example, in chemical reaction systems, the processes (that is, chemical reactions) are coupled through discrete reaction events that use molecules of some chemical species to generate others. Similarly, in the SIR model different processes influence each other by changing the state of the nodes, that is, from S to I in an infection event or from I to R in a recovery event. In the example shown in Fig. 2, when node v_4 recovers, it decreases the probability that its neighboring susceptible node v_1 gets infected within a certain time horizon compared with the scenario where v_4 remains infectious. As this example suggests, the probability that a susceptible node gets infected depends on the past states of its neighbors. Therefore, over the course of the entire simulation, the dynamics of a node's state (e.g., v_1) are dependent on those of its neighbors (e.g., v_2 and v_4).

Because of the coupling between jump processes, which is present in most systems of interest, we cannot simply simulate the system by separately generating the waiting times for each process according to Eq. (2.1). Any event that occurs will alter the processes to which it is coupled, thus rendering the waiting times we drew for the affected processes invalid. What the Gillespie algorithms do instead is to successively generate the waiting time until the next event, update the state of the system, and reiterate.

Besides being conditionally independent of each other, Poisson processes also display a temporal independence property, the so-called *memoryless property* (Box 1). In Poisson processes, the probability of the time to the next event, τ, is independent of how long we have already waited since the last event. In this sense, we do not need to worry about what has happened in the past. The only things that matter are the present status of the population (such as v_1 is susceptible and v_2 is infectious right now) and the model parameters (such as β and μ). The memoryless property can be seen as a direct consequence of the exponential distribution of waiting times of Poisson processes (Box 1, Eq.(2.2)).

2.6 Superposition of Poisson Processes

In this section, we explain a remarkable property of Poisson processes called the superposition theorem. The direct method exploits this theorem. Other methods, such as the rejection sampling algorithm (see Section 2.8) and the first reaction method, can also benefit from the superposition theorem to accelerate the simulations without impacting their accuracy.

Consider a susceptible individual v_i in the SIR model that is in contact with N_I infectious individuals. Any of the N_I infectious individuals may infect v_i. Consider the case shown in Fig. 3(a), where $N_I = 3$. If we focus on a single edge connecting v_i to one of its neighbors and ignore the other neighbors, the probability that v_i is infected via this edge exactly in time $[\tau, \tau + \delta t)$ from now, where δt is small, is given by $\psi_{S \to I}(\tau)\delta t$ (see Eq. (2.8)). Each of v_i's N_I neighbors may infect v_i in the same manner and independently. The neighbor that does infect v_i is the one for which the corresponding waiting time is the shortest, provided that it does not recover before it infects v_i. From this we can intuitively see that the larger N_I is, the shorter the waiting time before v_i gets infected tends to be. To simulate the dynamics of this small system, we need to know, not when each of its neighbors would infect v_i, but rather the time until any of its neighbors infects v_i.

To calculate the waiting-time distribution for the infection of v_i by any of its neighbors, we again resort to the discrete-time view of the infection processes. Because the infection processes are independent, the probability that v_i is not infected by any of its N_I infectious neighbors in a time window of duration δt is given by

$$(1 - \beta \delta t)^{N_{\mathrm{I}}}. \tag{2.9}$$

Therefore, the probability that v_i is infected after a time $\tau = r\delta t$ (i.e., v_i gets infected exactly in the $(r + 1)$th time window of length δt and not before) is given by

$$p_{\mathrm{I} \rightarrow \mathrm{R}} = \left[(1 - \beta \delta t)^{N_{\mathrm{I}}}\right]^r \times \left[1 - (1 - \beta \delta t)^{N_{\mathrm{I}}}\right]. \tag{2.10}$$

Here, the factor $\left[(1 - \beta \delta t)^{N_{\mathrm{I}}}\right]^r$ is the survival probability that an infection does not happen for a time $\tau = r\delta t$. The factor $\left[1 - (1 - \beta \delta t)^{N_{\mathrm{I}}}\right]$ is the probability that any of v_i's infectious neighbors infects v_i in the next time window, $t \in [\tau, \tau + \delta t)$.

Using the exponential identity $\lim_{x \rightarrow 0}(1 + x)^{1/x} = e$ with $x = -\beta \delta t$ as we did in Section 2.4, we obtain in the continuous-time limit that

$$\lim_{\delta t \rightarrow 0}\left[(1 - \beta \delta t)^{N_{\mathrm{I}}}\right]^r = \lim_{\delta t \rightarrow 0}\left[(1 - \beta \delta t)^{1/(-\beta \delta t)}\right]^{-N_{\mathrm{I}}\beta \tau} = e^{-N_{\mathrm{I}}\beta \tau}, \tag{2.11}$$

where the first equality is obtained by noting that $r = \tau/\delta t$ and rearranging the terms. In the same limit of $\delta t \rightarrow 0$, we obtain from Taylor expansion that

$$\left[1 - (1 - \beta \delta t)^{N_{\mathrm{I}}}\right] \approx 1 - (1 - N_{\mathrm{I}}\beta \delta t) = N_{\mathrm{I}}\beta \delta t. \tag{2.12}$$

By combining Eqs. (2.10), (2.11), and (2.12), we obtain $p_{\mathrm{I} \rightarrow \mathrm{R}} \approx N_{\mathrm{I}}\beta e^{-N_{\mathrm{I}}\beta \tau}\delta t$. Therefore, the probability density with which v_i gets infected at time τ is given by

$$\psi_{\mathrm{I} \rightarrow \mathrm{R}}(\tau) = N_{\mathrm{I}}\beta e^{-N_{\mathrm{I}}\beta \tau}, \tag{2.13}$$

that is, the exponential distribution with rate parameter $N_{\mathrm{I}}\beta$. By comparing Eqs. (2.8) and (2.13), we see that the effect of having N_{I} infectious neighbors (see Fig. 3(a) for the case of $N_{\mathrm{I}} = 3$) is the same as that of having just one infectious neighbor with an infection rate of $N_{\mathrm{I}}\beta$.

This is a convenient property of Poisson processes, known as the superposition theorem (see Box 1, Eq. (2.3) for the general theorem). To calculate how likely it is that a susceptible node v_i will be infected in time τ, one does not need to examine when the infection would happen or whether the infection happens for each of the infectious individuals contacting v_i. We are allowed to agglomerate all those effects into one infectious supernode as if the supernode infects v_i with rate $N_{\mathrm{I}}\beta$. We refer to such a superposed Poisson process that induces a particular state transition in the system (in the present case, the transition from the S to the I state for v_i) as a *reaction channel*, following the nomenclature in chemical reaction systems.

This interpretation remains valid even if v_i adjacent to other irrelevant individuals. In the network shown in Fig. 3(b), the susceptible node v_i has degree

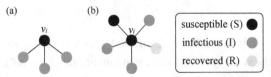

Figure 3 A susceptible node and other nodes surrounding it. (a) A susceptible node v_i surrounded by three infectious nodes. (b) A susceptible node v_i surrounded by five nodes in different states.

(i.e., number of other nodes that are connected to i by an edge) $k_i = 5$. Three neighbors of v_i are infectious, one is susceptible, and one is recovered. In this case, v_i will be infected at a rate of 3β, the same as in the case of v_i in the network shown in Fig. 3(a).

In both cases, we are replacing three instances of the probability density of the time to the next infection event, each given by $\beta e^{-\beta\tau}$, by a single probability density $3\beta e^{-3\beta\tau}$. Representing the three infectious nodes by one infectious supernode, that is, one reaction channel, with three times the infection rate is equivalent to superposing the three Poisson processes into one. Figure 4 illustrates this superposition, showing the putative event times generated by each Poisson process as well as those generated by their superposition. The superposition theorem dictates that the superposition is a Poisson process with a rate of 3β. This in particular means that we can draw the waiting time τ until the first of the events generated by all the three Poisson processes happens (shown by the double-headed arrow in Fig. 4) directly from the exponential distribution $\psi(\tau) = 3\beta e^{-3\beta\tau}$. Note that Poisson processes are defined as generating events indefinitely, and for illustrative purposes we show multiple events in Figure 4. However, in the SIR model only the first event in the superposed process will take place in practice. For example, once the event changes the state of v_i from S to I, the node cannot be infected anymore, and therefore none of the three infection processes can generate any more events.

Let us consider again the snapshot of the SIR dynamics shown in Fig. 3(a), but this time we consider all the possible infection and recovery events. We can represent all the possible events that may occur by four reaction channels (i.e., Poisson processes). One channel represents the infection of the node v_i by any of its neighbors, which happens at a rate 3β. We refer to this reaction channel as the first reaction channel. The three other channels each represent the recovery process of one of the infectious nodes. We refer to these three reaction channels as the second to the fourth reaction channels. We can use the same approach as above to obtain the probability density for the waiting time until the first event generated by any of the channels. However, to completely describe the dynamics, it is not sufficient to know when the next event happens. We also

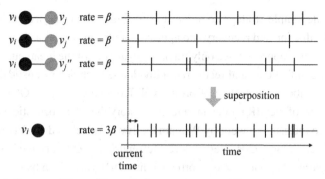

Figure 4 Superposition of three Poisson processes. The event sequence in the bottom is the superposition of the three event sequences corresponding to each of the three edges connecting v_i to its neighbors v_j, $v_{j'}$, and $v_{j''}$. The superposed event sequence generates an event whenever one of the three individual processes does. Note that edges (v_i, v_j), $(v_i, v_{j'})$, and $(v_i, v_{j''})$ generally carry different numbers of events in a given time window despite the rate of the processes (i.e., the infection rate, β) being the same. This is due to the stochastic nature of Poisson processes.

need to know which channel generates the event. Precisely speaking, we need to know the probability Π_i that it is the ith reaction channel that generates the event. Using the definition of conditional probability, we obtain

$$\Pi_i = \frac{\{\text{probability that an event in the } i\text{th reaction channel occurs}\}}{\{\text{probability that an event in any reaction channel } j \in \{1,2,3,4\} \text{ occurs}\}}.$$

$$(2.14)$$

In a discrete-time description, the numerator in Eq. (2.14) is simply $\lambda_i \delta t$, where $\lambda_1 = 3\beta$ and $\lambda_2 = \lambda_3 = \lambda_4 = \mu$ are the rates of the reaction channels. The denominator is equal to $1 - \prod_{j=1}^{4}(1 - \lambda_j \delta t)$, which in the limit of small δt can be Taylor expanded to $\sum_{j=1}^{4} \lambda_j \delta t = 3(\beta + \mu)\delta t$. Thus, the probability that the ith reaction channel has generated an event that has taken place is

$$\Pi_i = \frac{\lambda_i}{\sum_{j=1}^{4} \lambda_j}, \qquad (2.15)$$

that is, Π_i is simply proportional to the rate λ_i.

The same result holds true for general superpositions of Poisson processes (see Box 1).

2.7 Ignoring Stochasticity: Differential Equation Approach

We have introduced the types of models we are interested in and have explored their basic mathematical properties. We now turn our attention to the problem of how we can solve such models in practice. We consider again the SIR model. One simple strategy to solve it is to forget about the true stochastic nature of infection and recovery and approximate the processes as being deterministic. In this approach, we only track the dynamics of the mean numbers of susceptible, infectious, and recovered individuals. Such deterministic dynamics are described by a system of ordinary differential equations (ODEs). The ODE version of the SIR model has a longer history than the stochastic one, dating back to the seminal work by William Ogilvy Kermack and Anderson Gray McKendrick in the 1920s (Kermack & McKendrick, 1927). For the basic SIR model described previously, the corresponding ODEs are given by

$$\frac{d\rho_S}{dt} = -\beta\rho_S\rho_I, \tag{2.16}$$

$$\frac{d\rho_I}{dt} = \beta\rho_S\rho_I - \mu\rho_I, \tag{2.17}$$

$$\frac{d\rho_R}{dt} = \mu\rho_I, \tag{2.18}$$

where $\rho_S = N_S/N$, $\rho_I = N_I/N$, and $\rho_R = N_R/N$ are the fractions of S, I, and R individuals, respectively. The $\beta\rho_S\rho_I$ terms in Eqs. (2.16) and (2.17) represent infection events, through which the number of S individuals decreases and the number of I individuals increases by the same amount. The $\mu\rho_I$ terms in Eqs. (2.17) and (2.18) represent recovery events.

One can solve Eqs. (2.16), (2.17), and (2.18) either analytically, to some extent, or numerically using an ODE solver implemented in various programming languages. Suppose that we have coded up Eqs. (2.16), (2.17), and (2.18) into an ODE solver to simulate the infection dynamics (such as time courses of ρ_I) for various values of β and μ. Does the result give us complete understanding of the original stochastic SIR model? The answer is negative (Mollison, Isham, & Grenfell, 1994), at least for the following reasons.

First, the ODE is not a good approximation when N is small. In Eqs. (2.16), (2.17), and (2.18), the variables are the fraction of individuals in each state. For example, $\rho_I = N_I/N$. The ODE description assumes that ρ_I can take any real value between 0 and 1 and that ρ_I changes continuously as time goes by. However, in reality ρ_I is quantized, so it can take only the values $0, 1/N, 2/N,$ $\ldots (N-1)/N$, and 1, and it changes in steps of $1/N$ (e.g., it changes from $3/N$ to $4/N$ discontinuously). This discrete nature does not typically cause serious problems when N is large, in which case ρ_I is close to being continuous. By contrast, the ODE model is not accurate when N is small due to the quantization effect. (Note that the ODE approach is problematic in some cases even when N is large, i.e. near critical points, as we discuss in what follows.)

Second, even if N is large, the actual dynamic changes in ρ_I, for example, are not close to what the ODEs describe when the number of infectious individuals is small. For example, if $\rho_I = 2/N$, there are two infectious individuals. If one of them recovers, ρ_I changes to $1/N$, and this is a 50 percent decrease in ρ_I. The ODE assumes that ρ_I changes continuously and is not ready to describe such a change. As another example, suppose that we initially set $\rho_S = (N-1)/N$, $\rho_I = 1/N$, and $\rho_R = 0$. In other words, there is a single infectious seed, and all the other individuals are initially susceptible. In fact, the theory of the ODE version of the SIR model shows that ρ_I increases deterministically, at least initially, if $\beta > \mu$, corresponding to the situation in which an outbreak of infection happens. However, in the stochastic SIR model, the only initially infectious individual may recover before it infects anybody even if $\beta > \mu$. When this situation occurs, the dynamics terminate once the initially infectious individual has recovered, and no outbreak is observed. Although the probability with which this situation occurs decreases as β/μ increases, it is still not negligibly small for many large β/μ values. This is inconsistent with the prediction of the ODE model. It should be noted that another common way to initialize the system is to start with a small fraction of infectious individuals, regardless of N. In this case, if we start the stochastic SIR dynamics in a large, well-mixed population and, for example, with 10 percent initially infectious individuals, the ODE version is sufficiently accurate at describing the stochastic SIR dynamics.

Third, ODEs are not accurate at describing the counterpart stochastic dynamics when the system is close to a so-called critical point. For example, in the SIR model, given the value of the infection rate (i.e., μ), there is a value of the infection rate called the epidemic threshold, which we denote by β_c. For $\beta < \beta_c$, only a small number of the individuals will be infected (i.e., the final epidemic size is of $O(1)$). For $\beta > \beta_c$, the final epidemic size is large (i.e., $O(N)$) with a positive probability. In analogy with statistical physics, β_c is termed a critical point of the SIR model. Near criticality the fluctuations of ρ_S, ρ_I, and ρ_R are not negligible compared to their mean values, even for large N, and the ODE generally fails.

Fourth, ODEs are not accurate when dynamics are mainly driven by stochasticity rather than by the deterministic terms on the right-hand sides of the ODEs. This situation may happen even far from criticality or in a model that does not show critical dynamics. The voter model (see Section 4.10.3 for details) is such a case. In its simplest version, the voter model describes the tug-of-war between two equally strong opinions in a population of individuals. Because the two opinions are equally strong, the ODE version of the voter model predicts that the fraction of individuals supporting opinion A (and that of individuals supporting opinion B) does not vary over time, that is, one obtains $d\rho_A/dt = d\rho_B/dt = 0$, where ρ_A and ρ_B are the fractions of individuals supporting opinions A and B, respectively. However, in fact, the opinion of the individuals flips here and there in the population due to stochasticity, and it either increases or decreases over time.

To summarize, when stochasticity manifests itself, the approximation of the original stochastic dynamics by an ODE model is not accurate.

2.8 Rejection Sampling Algorithm

The most intuitive method to simulate the stochastic SIR model, while accounting for the stochastic nature of the model, is probably to discretize time and simulate the dynamics by testing whether each possible event takes place in each step. This is called the rejection sampling algorithm. Let us consider the stochastic SIR model on a small network composed of $N = 6$ nodes, as shown in Fig. 2, to explain the procedure.

Assume that the state of the network (i.e., the states of the individual nodes) is as shown in Fig. 2(a) at time t; three nodes are susceptible, two nodes are infectious, and the other node is recovered. In the next time-step, which accounts for a time length of Δt and corresponds to the time interval $[t, t + \Delta t)$, an infection event may happen in five ways: v_2 infects v_1, v_2 infects v_3, v_2 infects v_5, v_4 infects v_1, and v_4 infects v_5. Recovery events may happen for v_2 and v_4. Therefore, there are seven possible events in total, some of which may simultaneously happen in the next time-step.

With the rejection sampling method, we sequentially (called asynchronous updating) or simultaneously (called synchronous updating) check whether or not each of these events happens in each time-step of length Δt. Note that it is not possible to go to the limit of $\Delta t \to 0$ in rejection sampling. In our example, v_2 infects v_1 with probability $\beta \Delta t$ in a time-step. With probability $1 - \beta \Delta t$, nothing occurs along this edge. In practice, to determine whether the event takes place or not, we draw a random number u uniformly from $[0, 1)$. If $u \geq \beta \Delta t$, the algorithm rejects the proposed infection event (thus the name rejection sampling). If $u < \beta \Delta t$, we let the infection occur. Then, under asynchronous updating, we change the state of v_1 from S to I and update the set of possible events accordingly right away, and then proceed to check the occurrence of each of the remaining possible events in turn. Under synchronous updating, we first check whether each of the possible state changes takes place and note down the changes that take place. We then implement all the noted changes simultaneously. Regardless of whether we use asynchronous or synchronous updating, the infection event occurs with probability $\beta \Delta t$.

If v_4 recovers, which occurs with probability $\mu \Delta t$, and none of the other six possible events occurs in the same time-step, the status of the network at time $t + \Delta t$ is as shown in Fig. 2(b). Then, in the next time-step, v_1 may get infected, v_2

may recover, v_3 may get infected, and v_5 may get infected, which occurs with probabilities $\beta\Delta t$, $\mu\Delta t$, $\beta\Delta t$, and $\beta\Delta t$, respectively. In this manner, we carry forward the simulation by discrete steps until no infectious nodes are left.

There are several caveats to this approach. First, the asynchronous and the synchronous updating schemes of the same stochastic dynamics model may lead to systematically different results (Cornforth, Green, & Newth, 2005; Greil & Drossel, 2005; Huberman & Glance, 1993).

Second, one should set Δt such that both $\beta\Delta t < 1$ and $\mu\Delta t < 1$ always hold true. In fact, the discrete-time interpretation of the original model is justified only when Δt is small enough to yield $\beta\Delta t \ll 1$ and $\mu\Delta t \ll 1$.

Third, in the case of asynchronous updating, the order of checking the events is arbitrary, but it affects the outcome, particularly if Δt is not tiny. For example, we can sequentially check whether each of the five infection events occurs and then whether each of the two recovery events occurs, completing one time-step, One can alternatively check the recovery events first and then the infection events. If we do so and v_4 recovers in the time-step, then it is no longer possible that v_4 infects v_1 or v_5 in the same time-step because v_4 has recovered. If the infection events were checked before the recovery events, it is possible that v_4 infects v_1 or v_5 before v_4 recovers in the same time-step.

Fourth, some of the seven types of event cannot occur simultaneously in a single time-step regardless of whether the updating is asynchronous or synchronous, and regardless of the order in which we check the events in the asynchronous updating. For example, if v_2 has infected v_1, then v_4 cannot infect v_1 in the same time-step (or anytime later) and vice versa. In fact, from the susceptible node v_1's point of view, it does not matter which infectious neighbor, either v_2 or v_4, infects v_1. What is primarily important is whether v_1 gets infected or not in the given time-step, whereas one wants to know who infected whom in some tasks such as contact tracing.

A useful method to mitigate the effect of overlapping events of this type is to take a node-centric view. The superposition theorem implies that v_1 will get infected according to a Poisson process with rate 2β because it has two infectious neighbors (Section 2.6). By exploiting this observation, let us redefine the list of possible events at time t. The node v_1 will get infected with probability $2\beta\Delta t$ (and will not get infected with probability $1 - 2\beta\Delta t$). Nodes v_3 and v_5 will get infected with probabilities $\beta\Delta t$ and $2\beta\Delta t$, respectively. As before, v_2 and v_4 recover with probability $\mu\Delta t$ each. In this manner, we have reduced the number of possible events from seven to five. We are usually interested in simulating such stochastic processes in much larger networks or populations, where nodes tend to have a degree larger than in the network shown in Fig. 2. For example, if a node v_i has 50 infectious neighbors, implementing the rejection sampling using the probability that v_i gets infected, $50\beta\Delta t$, rather than checking if v_i gets infected with probability $\beta\Delta t$ along each edge that v_i has with an infectious neighbor, will confer a fiftyfold speedup of the algorithm.

Rejection sampling is a widely used method, particularly in research communities where continuous-time stochastic process thinking does not prevail. In a related vein, many people are confused by being told that the infection and recovery rates β and μ can exceed 1. They are accustomed to think in discrete time such that they are not trained to distinguish between the rate and probability. They *are* different; simply put, the rate is for continuous time, and the probability is for discrete time. Here we advocate that we should not use the discrete-time versions in general, despite their simplicity and their appeal to our intuition, for the following reasons (see Gómez, Gómez-Gardeñes, Moreno, and Arenas [2011] and Fennell, Melnik, and Gleeson [2016] for similar arguments).

First, the use of a small Δt, which is necessary to assure an accurate approximation of the actual continuous-time stochastic process, implies a large computation time. If the duration of time that one run of simulation needs is T, one needs $n = T/\Delta t$ discrete time-steps, which is large when Δt is small. How small should Δt be? It is difficult to say. If you run simulations with a choice of a small Δt and calculate statistics of your interest or draw a figure for your report or paper, a good practice is to try the same thing after halving Δt. If the results do not noticeably change, then your original choice of Δt is probably small enough for your purpose. Otherwise, you need to make Δt smaller. It takes time to carry out such a check just to determine an appropriate Δt value. Many people skip it. The Gillespie algorithms do not rely on a discrete-time approximation and are also typically faster than rejection sampling with a reasonably small Δt value.

Second, no matter how small Δt is, the results of rejection sampling are only approximate. This is because it is exact only in the limit $\Delta t \to 0$. By contrast, the Gillespie algorithms are always exact.

Proponents of the rejection sampling method may say that they want to define the model (such as the SIR model) in discrete time and run it rather than to consider the continuous-time version of the model and worry about the choice of Δt or the accuracy of the rejection sampling. We recommend against this as well. In the SIR model in discrete time, any infectious individual v_i infects a neighboring susceptible individual v_j with probability β', and each infectious individual recovers with probability μ'. Then, there are at least two problems related to this. First, the order of the events affects dynamics in the case of asynchronous updating. Second, and more importantly, we do not know how to change the time resolution of the simulation when we need to. For example, if one simulation step currently corresponds to one hour, one may want to now simulate the same model with some more temporal detail such that one step corresponds to ten minutes. Because the physical time is now one sixth of the original one, should we multiply β' and μ' by $1/6$ and do the same simulations? The answer is no. If the original time-step corresponds to $\Delta t = 1$, which is often implicit, then the probability that an infectious individual recovers in the continuous-time stochastic SIR model within time $\Delta t (= 1)$ is $1 - e^{-\mu \Delta t} = 1 - e^{-\mu}$, which we equate with μ'. Then, if we scale the time c times (e.g., $c = 1/6$), the probability that the recovery occurs in a new single time-step is $1 - e^{-\mu c \Delta t} = 1 - e^{-c\mu}$, which is not equal to $c\mu'$. For example, with $\mu = 1$ and $c = 1/6$, one obtains $1 - e^{-c\mu} \approx 0.154$, whereas $c\mu' \approx 0.105$.

3 Classic Gillespie Algorithms

The Gillespie algorithms overcome the two major drawbacks of the rejection sampling algorithm that we discussed near the end of Section 2.8; namely, its computational inefficiency and its reliance on a discrete-time approximation of the dynamics. The Gillespie algorithms are typically faster than rejection sampling, and they are stochastically exact (i.e., they generate exact realizations of the simulated jump processes). In this section, we present the two basic Gillespie algorithms for simulating coupled Poisson processes, largely in their original forms proposed by Daniel Gillespie: the *first reaction method* and the *direct method*. The two methods are mathematically equivalent (Gillespie, 1976). Nevertheless, the two algorithms have pros and cons in terms of ease of implementation and computational efficiency. Because these two factors depend on the model to be simulated, which algorithm one should select depends on the model as well as personal preference.

We first provide a brief history of the Gillespie algorithms (Section 3.1). We then introduce the first reaction method (Section 3.2) because it is conceptually the simpler of the two, followed by the direct method (Section 3.3), which builds on elements of the first reaction method but makes use of the superposition theorem (Box 1) to directly draw the waiting time between events. We end this section with example implementations in Python of the stochastic SIR dynamics.

3.1 Brief History

As the name suggests, the Gillespie algorithms are ascribed to American physicist Daniel Thomas Gillespie (Gillespie, 1976, 1977). He originally proposed them in 1976 for simulating stochastic chemical reaction systems, and they have seen many applications as well as further algorithmic developments in this field. Nevertheless, the algorithms only rely on general properties of Poisson processes and not on any particular properties of chemical reactions. Therefore, the applicability of the Gillespie algorithms is much wider than to chemical reaction systems. In fact, they have been extensively used in simulations of multiagent systems both in unstructured populations and on networks. The only assumptions are that the system undergoes changes via sequences of discrete events (e.g., somebody infects somebody, somebody changes its internal state from a low-activity state to a high-activity state) and with a rate that stays constant in between events. Nevertheless, the latter assumption has been relaxed in recent extensions of the algorithms; we review these in Section 5.

There were precursors to the Gillespie algorithms. The American mathematician Joseph Leo Doob developed in his 1942 and 1945 papers the mathematical foundations of continuous-time Markov chains that underlie the Gillespie algorithms (Doob, 1942, 1945). In the second of the two papers, he effectively proposed the direct method, although the focus of the paper was mathematical theory and he did not propose a computational implementation (Doob, 1945, pp. 465–466). Due to this, the algorithm is sometimes called the Doob–Gillespie algorithm. David George Kendall, who is famous for Kendall's notation in queuing theory,[4] implemented an equivalent of the direct method to simulate a stochastic birth-death process on a computer as early as in 1950 (Kendall, 1950). In 1953, Maurice Stevenson Bartlett, a British statistician, simulated the SIR model in a well-mixed population (i.e., every pair of individuals is directly connected to each other) using the direct method (Bartlett, 1953).

Independently of Gillespie, Alfred B. Bortz and colleagues also proposed the same algorithm as the direct method to simulate stochastic dynamics of Ising spin systems in statistical physics in 1975 (Bortz, Kalos, & Lebowitz, 1975). Therefore, the direct method is also called the *Bortz–Kalos–Lebowitz algorithm* (or the *n-fold way* following the naming in their paper, and also *rejection-free kinetic Monte Carlo* and the *residence-time algorithm*). An even earlier paper published in 1966 in the same field proposed almost the same algorithm, with the only difference being that the waiting time between events was assumed to take a deterministic value rather than being stochastic as in the Gillespie algorithms (Young & Elcock, 1966).

---

[4] Not to be confused with another British statistician of the time, Maurice Kendall, famous for Kendall tau rank correlation. Both Kendalls were awarded the honor of the Royal Statistical Society, the Guy Medal in Gold.

3.2 First reaction method

To introduce the first reaction method, we consider our earlier example of the SIR model on a six-node network (see Fig. 2(a)). Here two types of events may happen next: either a susceptible node becomes infected (S → I), or an infectious node recovers (I → R). The rates at which each node experiences a state transition are shown in Fig. 5(a), which replicates Fig. 2(a). For example, v_1 is twice as likely to be infected next as v_3 is because v_1 has two infectious neighbors whereas v_3 has one infectious neighbor. Each event obeys a separate Poisson process. Therefore, let us first generate hypothetical event sequences according to each Poisson process with their respective rates (see Fig. 5(b)). In fact, we need to use at most only the first event in each sequence (shown in magenta in Fig. 5(b)). For example, in the event sequence for v_1, the first event may be used, in which case v_1 will be infected. Once v_1 is infected, the subsequent events on the same event sequence will be discarded because v_1 will never be infected again. If v_1 is infected, it will undergo another type of event, which is recovery. However, we cannot reuse the second or any subsequent events in the same sequence as the recovery event because the recovery occurs with rate μ, which is different from the rate 2β with which we have generated the event sequence for v_1. A lesson we learn from this example is that we should not prepare many possible event times beforehand because most of them would be discarded.

Given this reasoning, Gillespie proposed the first reaction method in one of his two original papers on the Gillespie algorithms (Gillespie, 1976). The idea of the first reaction method is to generate only the first putative event time (shown in magenta in Fig. 5(b)) for each node. These putative times correspond to when each node would experience a next event if nothing else happens in the rest of the system. We do not generate an event time for v_6 because this node is recovered, so it will never undergo any event. We then figure out which event occurs first. (In the example in Fig. 5(a), it is node v_4 that will change its state, and the state change is from I to R.)

To generate a putative waiting time for each node $v_1, \ldots, v_5$ in practice, we use a general technique called *inverse sampling*, which proceeds as follows. For example, the time to the first event for v_1 obeys the exponential distribution $\psi_1(\tau) \equiv 2\beta e^{-2\beta\tau}$. The probability that the time to the next event is larger than τ, called the survival probability (also called the survival function and the complementary cumulative distribution function), is given by

$$\Psi_1(\tau) \equiv \int_\tau^\infty \psi_1(\tau')d\tau' = \int_\tau^\infty 2\beta e^{-2\beta\tau'} d\tau' = e^{-2\beta\tau}. \tag{3.1}$$

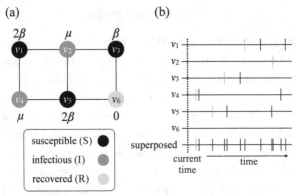

Figure 5 Determination of the time to the next event in the SIR model using the Gillespie algorithms. (a) The current state of the system. The rate with which each node changes its state is shown next to the node. (b) The putative events generated by the Poisson process corresponding to each node, with the first event of each process shown in magenta. There is no event on the timeline of v_6. This is because v_6 is in the recovered state and no longer undergoes any state change; therefore the event rate of the Poisson process associated with v_6 is equal to 0. The first reaction method generates a putative waiting time until the first event for each process and selects the smallest one. The direct method directly generates the waiting time until the first event in the superposed process (bottom of panel (b)).

By definition, $\Psi_1(\tau)$ is a probability, so $0 < \Psi_1(\tau) \leq 1$. We have excluded $\Psi_1(\tau) = 0$ because it happens only in the limit $\tau \to \infty$. This exclusion does not cause any problem in the following development.

Then, we draw a number u from an unbiased random number generator that generates random numbers uniformly in the interval $(0, 1]$. The generated number u is called a uniform $(0, 1]$ random variate. Any practical programming language has a function to generate uniform (pseudo) random variates. We do, however, advise against using a programming language's standard pseudo-random number generator, which is typically of poor quality. You should instead use one from a scientific programming library or code it yourself. We will discuss some good practices for pseudorandom number generation in Section 4.9. Given u, we then generate τ from the implicit equation

$$u = \Psi_1(\tau). \tag{3.2}$$

For example, if we draw $u = 0.3$, we can find the unique τ value that satisfies Eq. (3.2). This method for generating random variates obeying a given distribution is called inverse (transform) sampling (von Neumann, 1951). One can use this method for a general probability density function, $\psi(\tau)$, as long as one can calculate its survival function, $\Psi(\tau) = \int_\tau^\infty \psi(\tau')d\tau'$, like in Eq. (3.1). In the present case, by combining Eqs. (3.1) and (3.2), we obtain

$$u = e^{-2\beta\tau}, \tag{3.3}$$

which leads to

$$\tau = -\frac{\ln u}{2\beta}. \tag{3.4}$$

Note that $\tau \geq 0$ because $\ln u \leq 0$. Equation (3.4) is reasonable in the sense that a large event rate 2β will yield a small waiting time τ on average.

In the same manner, one can generate the putative times to the next event for $v_1, \ldots, v_5$, denoted by $\tau_1^{put}, \ldots, \tau_5^{put}$, using five uniform random variates. If the realized $\tau_1^{put}, \ldots, \tau_5^{put}$ values are as shown in Fig. 5(b), we conclude that node v_4 recovers next. Then, we change the state of v_4 from I to R and advance the clock by time $\tau = \tau_4^{put}$. Once v_4 recovers, the configuration of the six nodes will be the one given in Fig. 2(b). We then repeat the same procedure to find the next event given the updated set of processes corresponding to the nodes' states after the event (i.e., we now have three infection processes with rate β and one recovery process with rate μ), and so on.

Note that in our example, the event rate changed for v_1, v_4, and v_5, while it remained unchanged for v_2 and v_3. Therefore, we do not need to generate entirely new putative waiting times for v_2 and v_3. We just have to update τ_2^{put} and τ_3^{put} as $\tau_2^{put} \rightarrow \tau_2^{put} - \tau$ and $\tau_3^{put} \rightarrow \tau_3^{put} - \tau$, respectively, to account for the time τ that has elapsed.

In this manner, we can reuse τ_2^{put} and τ_3^{put} (by subtracting τ) and avoid having to generate new pseudorandom numbers for redrawing τ_2^{put} and τ_3^{put}. The effect of this frugality becomes important for larger systems where an event generally affects only a small fraction of the processes. The so-called *next reaction method* (Gibson & Bruck, 2000) exploits this idea to improve the computational efficiency of the first reaction method. Although the classic first reaction method did not make use of this trick, we include it here because it is simple to implement.

Let us go back to our example. For v_4, we no longer need to generate the time to the next event because v_4 is now in the R state. For v_1 and v_5, we need to discard τ_1^{put} and τ_5^{put} because they were generated under the assumption that the event rate was 2β. Now, we need to redraw the time to the next event for the two nodes according to the new distribution $\psi_1(\tau) = \psi_5(\tau) = \beta e^{-\beta\tau}$. For example, we reset $\tau_1^{put} = -\ln u'/\beta$, where u' is a new uniform $(0, 1]$ random variate. Although the time τ has passed to transit from the status of the network shown in Fig. 2(a) to that shown in Fig. 2(b), we do not need to take into account the elapsed time (i.e., τ) when generating the new τ_1^{put} and τ_5^{put} values. This is

Box 2 Gillespie's First Reaction Method

0. Initialization:

   (a) Define the initial state of system, and set $t = 0$.

   (b) Calculate the rate λ_j for each reaction channel $j \in \{1, \ldots, M\}$.

   (c) Draw M random variates u_j from a uniform distribution on $(0, 1]$.

   (d) Generate a putative waiting time $\tau_j^{\text{put}} = -\ln u_j / \lambda_j$ for each reaction channel.

1. Select the reaction channel i with the smallest τ_i^{put}, and set $\tau = \tau_i^{\text{put}}$.

2. Perform the event on reaction channel i.

3. Advance the time according to $t \to t + \tau$.

4. Update λ_i and all other λ_j that are affected by the event produced.

5. Update putative waiting times:

   (a) Draw new waiting times for reaction channel i and for the other reaction channels j whose λ_j has changed, according to $\tau_j^{\text{put}} = -\ln u_j / \lambda_j$ with u_j being newly drawn from a uniform distribution on $(0, 1]$.

   (b) Update the waiting times for all reaction channels j that have not been affected by the last event according to $\tau_j^{\text{put}} \to \tau_j^{\text{put}} - \tau$.

6. Return to Step 1.

due to the memoryless property of Poisson processes (see Box 1), that is, what happened in the past, such as how much time has passed to realize the state transition of v_4, is irrelevant.

The first reaction method in its general form is given in Box 2.

3.3 Direct Method

The direct method exploits the superposition theorem to directly generate the waiting times between successive events in the full system of coupled Poisson processes. For expository purposes, we hypothetically generate an event sequence on each node with the respective rate, although only at most the first event in each sequence will be used. We then superpose the nodal event sequences into one Poisson process (see Fig. 5(b)). Owing to the superposition theorem (Box 1, Eq. (2.3)), the superposed event sequence is itself a Poisson process with a rate Λ that is equal to the sum of the individual rates, that is, $\Lambda = 2\beta + \mu + \beta + \mu + 2\beta + 0 = 5\beta + 2\mu$.

Therefore, we can generate the time to the next event in the entire population using the inverse sampling method, which we introduced in Section 3.2, according to $\tau = -\ln u / \Lambda$, where u is a uniform random variate in the interval

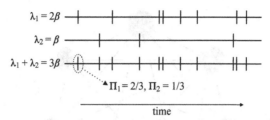

Figure 6 Superposition of Poisson processes and how to determine which component Poisson processes contribute to an event in the superposed event sequence. We consider $M = 2$ Poisson processes, one with rate $\lambda_1 = 2\beta$ and the other with $\lambda_2 = \beta$. The superposed Poisson process has rate $\lambda_1 + \lambda_2 = 3\beta$. The next event in the superposed event sequence (shown in the dotted circle) belongs to process 1 with probability $\Pi_1 = \lambda_1/(\lambda_1 + \lambda_2) = 2/3$ and process 2 with probability $\Pi_2 = \lambda_2/(\lambda_1 + \lambda_2) = 1/3$.

$(0, 1]$. We now know the time to the next event but not which node (or more generally, which one individual Poisson process) is responsible for the next event. This is because the superposition lacks information about the individual constituent event sequences.

We thus need to determine which node produces the event. The mathematical properties of Poisson processes guarantee that the probability that a given node produces the event is proportional to its event rate (see Box 1, Eq. (2.4)). For example, in Fig. 6, where the event rates of the two nodes v'_1 and v'_2 are 2β and β, respectively, each event in the superposed event sequence comes from v'_1 and v'_2 with probability $\Pi_1 = 2/3$ and $\Pi_2 = 1/3$, respectively. Therefore, the probability that the first event is generated by v'_1 is $2/3$. The probability it is generated by v'_2 is $1/3$. This is natural because the sequence for v'_1 has on average twice as many events as that for v'_2. In our example (see Fig. 5), the v_i (with $i = 1, \ldots, 5$) that produces the next event is drawn with probability Π_i, where $\Pi_1 = \Pi_5 = 2\beta/(5\beta+2\mu)$, $\Pi_2 = \Pi_4 = \mu/(5\beta+2\mu)$, and $\Pi_3 = \beta/(5\beta+2\mu)$. We will explain computational methods for doing this later.

Assume that v_4 generates the next event and transitions from state I to state R. Then, the new event rate for each node is as shown in Fig. 2(b). We advance the clock by τ and go to the next step. Again, to determine the time to the following event, regardless of which node produces the event, we only need to consider the sum of the event rates of the six nodes, which is now given by $\Lambda = \beta + \mu + \beta + 0 + \beta + 0 = 3\beta + \mu$. Then, the time to the next event, which we again denote by τ, is given by $\tau = -\ln u/(3\beta + \mu)$, where u is a new uniform random variate. We draw another random number to determine which of the four eligible nodes, v_1, v_2, v_3, or v_5, produces the event and changes its state. (Note that v_4 and v_6 are recovered, so they cannot undergo a further state

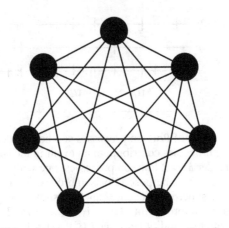

Figure 7 Complete graph with $N = 7$ nodes.

change.) The four remaining nodes are each selected with probability $\Pi_1 = \Pi_3 = \Pi_5 = \beta/(3\beta + \mu)$ and $\Pi_2 = \mu/(3\beta + \mu)$.

In this manner, we draw τ, determine which node produces the event, implement the state change, advance the clock by τ, and repeat. This is Gillespie's direct method.

Let us take a look at another example, which is the SIR model in a population composed of N individuals, in which everybody is directly connected to anybody else (called a well-mixed population; equivalent to a complete graph; see Fig. 7 for an example). In contrast to the general network case, each individual in a well-mixed population is indistinguishable from the others. In the well-mixed population, it is not the case that, for example, an individual has two neighbors while another has three neighbors (such as in Fig. 2); they all have $N - 1$ neighbors. Recall that recovered individuals do not change their state again. Therefore, it suffices to consider $N_S + N_I$ event sequences and their superposition to determine the time to the next event and which individual will produce the next event.

However, in the well-mixed population, we can make the procedure more efficient. Because everybody is alike, we do not need to keep track of the state of each individual.

In the case of a network, we generally need to distinguish between different susceptible individuals. For example, in Fig. 2(a), the susceptible nodes v_1 and v_5 are different because they have different sets of neighbors. Furthermore, v_1 has degree 2, while v_5 has degree 3. Therefore, if v_1 gets infected and changes its state, it is not equivalent to v_5 getting infected. So, we must keep track of the state of each individual in a general network. By contrast, in the well-mixed population, such a distinction is irrelevant. Everybody is adjacent to all the other $N-1$ nodes, and the number of infectious neighbors is the same for any susceptible individual, that is, it is equal to N_I. It does not matter which particular node gets infected in the next event. The only thing that matters for the SIR model in the well-mixed population is the number of susceptible, infectious, and recovered individuals, which is N_S, N_I, and N_R, respectively.

It thus suffices to monitor these numbers. If an infection event happens, then N_S decreases by one, and N_I increases by one. If an infectious individual recovers, then N_I decreases by one, and N_R increases by one. Because the number of individuals is preserved over time, it holds true that

$$N_S + N_I + N_R = N \tag{3.5}$$

at any time. Because anybody is connected to everybody else, any susceptible individual has N_I infectious neighbors, so it gets infected at rate βN_I. Because recovery occurs independently of a node's neighbors, every infectious individual recovers at the same rate μ.

Based on this reasoning, we can aggregate the event sequences of the N_S susceptible individuals into one even before considering which method we should use to simulate the SIR dynamics. (Therefore, this logic also works for the first reaction method and for rejection sampling.) Each event sequence corresponding to a single susceptible individual has the associated event rate $N_I \beta$. The superposed event sequence is a realization of a single Poisson process with rate $N_S \times N_I \beta = N_S N_I \beta$. If an event from this Poisson process occurs, one arbitrary susceptible individual gets infected. Likewise, we do not need to differentiate between the N_I infectious individuals. So, we superpose the N_I event sequences, each of which has the associated event rate μ, into an event sequence, which is a realization of a Poisson process with rate $N_I \times \mu$. If an event from this Poisson process occurs, then an arbitrary infectious individual recovers.

In summary, in a well-mixed population we only need to consider two coupled Poisson processes, one corresponding to contracting infection at a rate $\beta N_S N_I$, and the other corresponding to recovery at a rate μN_I. In a single step of the direct method, we first determine the waiting time to the next event, τ. We set $\tau = -\ln u/(\beta N_S N_I + \mu N_I)$, where u is a uniformly random $(0, 1]$ variate. Next, we determine which type of event happens, either the infection of a susceptible node (with probability $\Pi_{S \to I}$) or the recovery of an infectious node (with probability $\Pi_{I \to R}$). We obtain

$$\Pi_{S \to I} = \frac{\beta N_S N_I}{\beta N_S N_I + \mu N_I} = \frac{\beta N_S}{\beta N_S + \mu}, \tag{3.6}$$

$$\Pi_{I \to R} = \frac{\mu N_I}{\beta N_S N_I + \mu N_I} = \frac{\mu}{\beta N_S + \mu}. \tag{3.7}$$

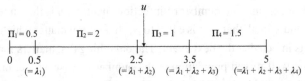

Figure 8 Linear search method for computing which Poisson process produces the next event for the entire population. We consider $M = 4$ possible events, with respective rates $\lambda_1 = 0.5$, $\lambda_2 = 2$, $\lambda_3 = 1$, and $\lambda_4 = 1.5$. Suppose that we draw a uniform random variate ranging between 0 and $\sum_{i=1}^{M} \lambda_i = 5$ whose value is $u = 2.625$. We first check whether u falls inside the first interval; in practice, we check if $u \leq \lambda_1 = 0.5$. Because this is not the case, we then check iteratively if it lies in each following interval. Because $2.5 < u \leq 3.5$, we find that u falls in the third interval from the left (dotted arrow). The iteration over i as described in Step 2 in Box 3 thus stops in the third interval, and the method will select $i = 3$.

If an infection event occurs, we decrease N_S by 1 and increase N_I by 1. If a recovery event occurs, we decrease N_I by 1 and increase N_R by 1. In either case, we advance the clock by τ and go to the next step. We repeat the loop until N_I hits 0.

In general applications of the direct method, we consider a set of M independent Poisson processes with rates λ_i ($1 \leq i \leq M$). The superposition of the M Poisson processes is a single Poisson process with rate $\Lambda = \sum_{i=1}^{M} \lambda_i$ by the superposition theorem. Therefore, the time to the next event in the entire population, τ, follows the exponential distribution given by

$$\psi(\tau) = \Lambda e^{-\Lambda \tau}. \tag{3.8}$$

After time τ, the ith process produces the next event with probability

$$\Pi_i = \frac{\lambda_i}{\Lambda}. \tag{3.9}$$

By drawing a random number obeying the categorical distribution over the M possibilities given by $\{\Pi_1, \ldots, \Pi_M\}$, we can then determine which Poisson process i generates one event. Gillespie's original implementation does this by iterating over the list of Π_i values (see Fig. 8).

We summarize the steps of the direct method in Box 3. Similarly to the first reaction method, the direct method is easy to implement, but it is not very fast in its original form (see Fig. 8) when M is large. For this reason more efficient algorithms have been proposed. We review them in Section 4.

Although we have assumed in our example that a reaction channel (i.e., a Poisson process in the present case) is attached to each node/individual, this does not always have to be the case. As we have seen, in the case of the

---

Box 3 Gillespie's Direct Method

0. Initialization:

    (a) Define the initial state of the system, and set $t = 0$.

    (b) Calculate the rate λ_j for each reaction channel $j \in \{1, \ldots, M\}$..

1. Draw a random variate u_1 from a uniform distribution on $(0, 1]$, and generate the waiting time by $\tau = -\ln u_1 / \Lambda$, where $\Lambda = \sum_{j=1}^{M} \lambda_j$ is the total rate.

2. Draw u_2 from a uniform distribution on $(0, \Lambda]$. Select the event i to occur by iterating over $i = 1, 2, \ldots, M$ until we find the i for which $\sum_{j=1}^{i-1} \lambda_j < u_2 \leq \sum_{j=1}^{i} \lambda_j$.

3. Perform the event on reaction channel i.

4. Advance the time according to $t \to t + \tau$.

5. Update λ_i as well as all other λ_j that are affected by the produced event.

6. Return to Step 1.

---

well-mixed population, we only need to track two reaction channels, that is, the number of susceptible individuals N_S and the number of infectious individuals N_I. In a more complicated setting where the network structure changes in addition to the nodes' states, some reaction channels are assigned to nodes, and other reaction channels may be assigned to the state of the edges, which may switch between on (i.e., edge available) and off (edge unavailable, or only weakly available) (Clementi et al., 2008; Fonseca dos Reis, Li, & Masuda, 2020; Kiss et al., 2012; Ogura & Preciado, 2016; Volz & Meyers, 2007; Zhang, Moore, & Newman, 2017). In all of the preceding cases, the key assumptions are that all types of reaction channels that trigger events are Poisson processes and that their event rate may only change in response to events generated by Poisson processes occurring anywhere in the population/network.

3.4 Codes

Here we present Python codes for the two classic Gillespie algorithms for simulating the SIR model. Then, we compare their output and runtimes to each other and to the rejection sampling algorithm presented in Section 2.8. We first show codes for the SIR model in a well-mixed population (Section 3.4.1). We then present implementations for the SIR model on a network (Section 3.4.2), which requires additional bookkeeping to track the dependencies between nodes and the varying number of reaction channels.

All our example codes rely on the NumPy library in Python for vectorized computation and for generating pseudorandom numbers. The following code imports the NumPy library and initializes the pseudorandom number generator (see Section 4.9 for a discussion of how to generate random numbers on a computer).

```
import numpy as np
from numpy.random import Generator, PCG64

seed = 42                        # Set seed for PRNG state
rg = Generator(PCG64(seed))   # Initialize random number
    generator
```

The following codes as well as those for the rejection sampling algorithm and those for producing figures are found in Jupyter notebooks at github.com/naokimas/gillespie-tutorial.

3.4.1 SIR Model in Well-Mixed Populations

The major part of the code for simulating the SIR model is identical for the first reaction and direct methods and is simply related to updating and saving the system's state. We thus show only the code needed for generating the waiting time and the next event in each iteration.

The following code snippet implements the first reaction method for the SIR model in a well-mixed population:

```
def draw_next_event_first_reaction(lambda_inf, lambda_rec):
    '''Input: total infection and recovery rates, lambda_inf=S*I
    *beta_k and lambda_rec=I*mu, respectively.
    Output: selected reaction channel, i_selected, and waiting
    time until the event, tau.'''

    # Draw a uniform random variate from (0,1] for each waiting
    time:
    u = 1. - rg.random(2)

    # Draw waiting times:
    waiting_times = - np.log(u) / np.array([lambda_inf,
    lambda_rec])

    # Select reaction with minimal tau:
    tau        = np.min(waiting_times)
    i_selected = np.argmin(waiting_times)

    return(i_selected, tau)
```

The following snippet implements the direct method:

```
def draw_next_event_direct(a_inf, a_rec):
    '''Input: total infection and recovery rates, lambda_inf=S*I
    *beta_k and lambda_rec=I*mu, respectively.
    Output: selected reaction channel, i_selected, and the
    waiting time until the event, tau.'''

    # Calculate cumulative rate:
    Lambda = lambda_inf + lambda_rec

    # Draw two uniform random variates from (0,1]:
```

```
 9    u1, u2 = 1. - rg.random(2)
10
11    # Draw waiting time:
12    tau = - np.log(u1) / Lambda
13
14    # Select reaction and update state:
15    if u2 * Lambda < lambda_inf: # S->I reaction
16        i_selected = 0
17    else:                        # I->R reaction
18        i_selected = 1
19
20    return(i_selected, tau)
```

Finally, the following code snippet, which is common to the two methods, implements the state update after the next event has been selected and recalculates the values of the infection and recovery rates:

```
1  # Update state:
2  if i_selected == 0: # S->I reaction
3      S -= 1; I += 1
4  else:               # I->R reaction
5      I -= 1 ;R += 1
6
7  # Update infection and recovery rates:
8  lambda_inf = S * I * beta_k # Infection rate
9  lambda_rec = I * mu         # Recovery rate
```

We compare simulation results for the SIR model in a well-mixed population with $N = 100$ individuals among the rejection sampling, Gillespie's first reaction method, and Gillespie's direct method in Fig. 9. The time-step used for rejection sampling shown in Fig. 9(a) is $\Delta t = 0.1$ With this time-step, rejection sampling leads to an undershoot of the peak number of infectious individuals; compare Fig. 9(a) to Figs. 9(b) and 9(c). We also note that the average curves, shown by the solid lines, fail to capture the large variation and bimodal nature of the stochastic SIR dynamics in all the panels. Finally, we note that the average runtimes of the three different algorithms to generate one simulation are of the order of 300 ms for the rejection sampling algorithm and of the order of 10 ms for both the Gillespie algorithms.

3.4.2 SIR Model on a Network

To simulate the SIR model on a network, we rely on the NetworkX library in addition to NumPy to store and update information about nodes in the network as well as their event rates. We import NetworkX as follows:

```
1  import networkx as nx
```

The following code implements the generation of a single event of the SIR model on a network G using the first reaction method. Here, G stores the connections between nodes as well as each node's state (S, I, or R), event rate, and putative waiting time.

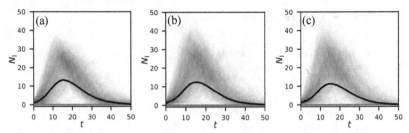

Figure 9 Evolution of the number of infectious individuals N_I over time of the SIR model in a well-mixed population simulated using (a) the rejection sampling method, (b) the first reaction method, and (c) the direct method. The population is composed of $N = 100$ individuals. The infection rate is $\beta = 0.5$. The recovery rate is $\mu = 0.2$. We carried out 1 000 simulations with each method. Each panel shows the number of infectious individuals over time. The overlapping thin red lines show the result for each of the 1 000 simulations. The thick black lines show the average over the 1 000 simulations. Note that a substantial portion of the simulations do not lead to any secondary infections; the red lines drop rapidly to zero, appearing as straight red lines at $N_I = 0$. Otherwise, N_I increases first and then decays towards zero. The average behavior does not capture this bimodal nature of the stochastic dynamics.

```
1  def draw_next_event_first_reaction(G):
2      '''Input: the network G.
3      Output: selected reaction channel, i_selected, and the
       waiting time until the event, tau.'''
4
5      # Get waiting times for active reaction channels from G:
6      node_indices  = list(nx.get_node_attributes(G, 'tau'))
7      waiting_times = list(nx.get_node_attributes(G, 'tau').values
       ())
8
9      # Select reaction with minimal waiting time:
10     tau = np.min(waiting_times)
11
12     i_selected = np.where(waiting_times == tau)[0][0]
13
14     return(i_selected, tau)
```

The following code implements the generation of a single event of the SIR model on a network G using the direct method. Here, G stores the connections between nodes as well as each node's state (S, I, or R) and its event rate.

```
1  def draw_next_event_direct(Lambda, G):
2      '''Input: the network, G, and the total event
3      rate, Lambda.
4      Output: selected reaction channel, i_selected, and the
       waiting time until the event, tau.'''
5
6      # Draw two uniform random variates from (0,1]:
7      u1, u2 = rg.random(2)
8
9      # Draw waiting time:
10     tau = - np.log(1. - u1) / Lambda
```

```
11
12      # Select reaction by linear search:
13      target_sum = u2 * Lambda
14      sum_i = 0
15
16      for i,attributes in G.nodes(data=True):
17          sum_i += attributes['lambda']
18
19          if sum_i >= target_sum:
20              break
21
22      return(i, tau)
```

Lines 13–20 implement the selection of the reaction channel that generates the next event using the simple algorithm illustrated in Fig. 8.

4 Computational Complexity and Efficient Implementations

In this section we investigate the computational efficiency of the Gillespie algorithms. We also review improvements that have been developed to make the algorithms more efficient when simulating systems with a large number of reaction channels.

A typical way to quantify the efficiency of stochastic algorithms, and the one we shall be concerned with here, is their *expected time complexity*. In the context of event-based simulations, it measures how an algorithm's average runtime depends on the number of reaction channels, M. While it is generally impossible to exactly calculate the expected runtime of an algorithm for all different use cases, we are often able to show how the algorithm's average runtime scales with M for large M. We indicate the complexity of an algorithm using *big-O notation*,[5] where $O(f(M))$ means that the algorithm's expected runtime is proportional to $f(M)$ for large M. For example, an algorithm with expected runtime $T_1(M) = 7M + 10$ and another with expected runtime $T_2(M) = 0.5M + \log M$ both have linear time complexity, that is, $T_i(M) = O(M)$ for $i = 1, 2$.

As we shall see in Section 4.1, the classic implementations of the Gillespie algorithms presented in Section 3 have $O(M)$ time complexity for each simulation step of the algorithms. While a linear time complexity may not seem computationally expensive, it is typical that the number of events taking place per time unit also scales linearly with the size of the system, namely N in most

---

[5] In the computer science literature, the big-O notation is often informally used to denote the expected time complexity of an algorithm. This differs from the formal definition of the big-O notation, which is pertinent to the worst-case complexity (Knuth, 1976). To keep things simple and to keep our notation consistent with the literature, we also adopt the big-O notation to denote the average time complexity.

of our examples. Thus, the number of computations per simulated unit of time then scales as $O(NM)$. This means that the overall time complexity of running a single simulation using the classic Gillespie algorithms is $O(NMT)$, where T is the typical duration of a single simulation. An $O(NMT)$ computation time may be prohibitively expensive for large N and M. Because M scales linearly with N at least, which occurs for sparse networks (i.e., networks with relatively few edges), we have at least $O(N^2T)$ time complexity in this scenario.

To make the Gillespie algorithms more efficient for large systems, researchers have come up with many algorithmic improvements to lower the computational complexity of both the bookkeeping and simulation steps of the algorithms. While these improved algorithms are more complex than the simple Gillespie algorithms presented in Section 3, many of these techniques are worth the effort to learn because they usually shorten the computation time immensely without sacrificing the exactness of the simulations. With these techniques, we may be able to simulate a large system that we could not simulate otherwise. However, we should note that, for systems with a small number of reaction channels, these methods will not confer a significant speedup and may even be slightly slower because they introduce some additional overhead. In this section we review several general methods that we believe to be the most important ones to be aware of for researchers looking to simulate social systems, although many more exist (see Marchetti, Priami, and Thanh [2017] for a recent review).

We detail in Section 4.1 the computational complexity of each step of the original first reaction and direct methods. We next discuss in Section 4.2 a simple way to improve the computational efficiency of the direct method by grouping similar processes together to reduce the number of reaction channels. We then review algorithmic improvements that decrease the expected complexity of both the direct (Section 4.3) and first reaction (Section 4.4) methods to $O(\log M)$ time. More recent methods further decrease the expected runtime of the direct method, which we review in Sections 4.5, 4.6, and 4.7. Other methods have been developed that sacrifice the exactness of the Gillespie algorithms to some extent for additional speed gains. Such methods are not the main focus of this Element, but we briefly review one such method, the *tau-leaping method*, in Section 4.8. We provide a short note on how to generate pseudorandom numbers needed for stochastic numerical simulations in Section 4.9. We end this section with example codes and simulation results (Section 4.10).

4.1 Average Complexity of the Classic Gillespie Algorithms

In this subsection we will analyze the runtime complexity of each step in Gillespie's two algorithms. Knowing which parts of the algorithms are the most computationally expensive will also tell us which parts of the algorithms we should focus on to make them more efficient.

We first analyze the complexity of the steps of the first reaction method in the order they appear in Box 2:

- **Step 1:** To find the smallest putative waiting time, we go through the entire list $\{\tau_1^{\text{put}}, \ldots, \tau_M^{\text{put}}\}$. Because there are M elements in the list, this step requires $O(M)$ time.

- **Steps 2–4:** Updating the system's state and event rates following an event requires a number of operations that is proportional to the number of reaction channels that are affected by the event.[6] In the case of a network, the number is often proportional to the average node degree. Typically the average node degree is relatively small and does not grow much with the size of the system, which we conventionally identify with the number of nodes. Such a network is referred to as a sparse network. So, this step takes $O(1)$ time. (However, for dense or heterogeneous networks, the number of reaction channels affected by an event may scale with M, in which case this step may take $O(M)$ time – see Section 4.7.)

- **Step 5:** To update the putative waiting times, we only have to generate new random variates for the reaction channel that generated the event and for those that changed their event rate due to the event. However, all the other waiting times still need to be updated in Step 5(b) in Box 2. Therefore, this step also has $O(M)$ time complexity.

We observe that, although we have avoided some costly parts of the original implementation of the first reaction method, which we introduced in Section 3.2, the algorithm still has linear time complexity. Thus, for systems with large M, the first reaction method may be slow.

Let us similarly analyze the time complexity per iteration of the direct method step-by-step in the order they appear in Box 3.

- **Step 1:** Generating the waiting time requires generating a random variate and transforming it. This is a constant-time operation, $O(1)$, because the runtime does not depend on M.

- **Step 2:** To find the reaction channel i that generates the event, we have to iterate through half of the list of the event rates on average. This step thus takes $O(M)$ time.

---

[6] Note that the original implementation of the first reaction method updates all λ_i at each timestep, which is an $O(M)$-time operation. However, we only need to update the rates that are affected by the event i, which is an $O(1)$-time operation if the average number of affected reaction channels is of constant order, that is, $O(1)$.

- **Steps 3–5:** The steps for updating the system's state and event rates are identical to Steps 2–4 for the first reaction method. Therefore, these steps typically have $O(1)$ time complexity.

Our implementation of the first reaction method has two steps of linear time complexity, whereas the direct method only has a single step of linear time complexity. Therefore, the direct method may be slightly faster than the first reaction method. However, the former's overall scaling with M is still linear. So, the direct method may be slow for large M, just like the first reaction method.

4.2 Grouping Reaction Channels

A simple way to reduce the effective number of reaction channels in the direct method, and thus accelerate the sampling of the next event, is available when the rates λ_i only take a small number of different values. The strategy is to group i's that have the same λ_i value and then apply a rounding operation, which is fast for a computer, to determine a unique value of i to be selected (Schulze, 2002).

This method works as follows. Consider an idealized situation in the SIR dynamics on a network in which $N = 100$, $N_S = 60$, $N_I = 30$, and $N_R = 10$. Let us further assume that, at the present moment in time, 40 out of the 60 susceptible individuals are adjacent to three infectious nodes, and the other 20 are adjacent to two infectious nodes (Fig. 10). We can safely ignore the $N_R = 10$ recovered individuals because they do not generate an event. For each susceptible individual with three and two infectious neighbors, the rate to get infected is 3β and 2β, respectively. For each infectious individual, the recovery rate is μ. Therefore, the total (i.e., cumulative) event rate is equal to $\Lambda = 40 \times 3\beta + 20 \times 2\beta + 30 \times \mu = 160\beta + 30\mu$. If the population is well-mixed and one just wants to track the numbers of the susceptible, infectious, and recovered individuals, one needs to prepare only two reaction channels, one for infection

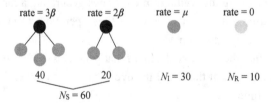

Figure 10 Diagram of the state of each node undergoing the SIR dynamics. Out of the $N = 100$ nodes, 40 nodes are susceptible and have three infectious neighbors (and other susceptible or recovered neighbors), 20 nodes are susceptible and have two infectious neighbors, 30 nodes are infectious, and 10 nodes are recovered.

(i.e., N_S decreases by 1, and N_I increases by 1) with rate 160β, and the other for recovery (i.e., N_I decreases by 1, and N_R increases by 1) with rate 30μ. However, the population that we are considering is not well-mixed because the number of infectious neighbors that a susceptible individual has at each moment in time depends explicitly on the states of its neighbors in the network. We thus need to keep track of each individual's state individually to be able to simulate the dynamics.

Each susceptible individual will generate the next event with probability

$$\Pi_i = \frac{3\beta}{160\beta + 30\mu} \tag{4.1}$$

or

$$\Pi_i = \frac{2\beta}{160\beta + 30\mu}, \tag{4.2}$$

depending on whether it has three or two infectious neighbors, respectively, while each infectious individual will generate the event with probability

$$\Pi_i = \frac{\mu}{160\beta + 30\mu}. \tag{4.3}$$

We group together the 40 susceptible individuals with three infectious neighbors, which altogether have a total rate of $\lambda'_1 \equiv 40 \times 3\beta = 120\beta$. Likewise, we group together the 20 susceptible individuals with two infectious neighbors, whose total rate is $\lambda'_2 \equiv 20 \times 2\beta = 40\beta$. The group of infectious individuals finally has a total rate of $\lambda'_3 \equiv 30 \times \mu = 30\mu$. Then, we determine which group is responsible for the next event. Because there are only three groups, this is computationally easy. In other words, we draw u_2 from a uniform distribution on $[0, 160\beta + 30\mu)$, and if $u_2 < \lambda'_1$, then it is group 1; if $\lambda'_1 \le u_2 < \lambda'_1 + \lambda'_2$, then it is group 2; otherwise, it is group 3.

We can next easily determine which individual in the selected group experiences the event. If group 1 is selected, we need to select one from the 40 susceptible nodes. Because their event rate is the same (i.e., $= 3\beta$), they all have the same probability to be selected. Therefore, one can select the ith individual (with $i = 1, \ldots, 40$) according to

$$i = \left\lfloor \frac{u_2}{\lambda'_1} \times 40 \right\rfloor + 1, \tag{4.4}$$

where $\lfloor \cdot \rfloor$ denotes rounding down to the nearest integer. Note that, because group 1 was selected, we have conditioned on $0 \le u_2 < \lambda'_1$, and u_2/λ'_1 is thus a uniform random variate on $[0, 1)$. Therefore, $u_2/\lambda'_1 \times 40$ is uniformly randomly distributed on $[0, 40)$. By rounding down this number, we can sample each integer from 0 to 39 with equal probability, namely $1/40$. The term $+1$ on the right-hand side of Eq. (4.4) lifts the sampled integer by one to guarantee that i is an integer between 1 and 40. Likewise, if group 2 has been selected, we set

$$i = \left\lfloor \frac{u_2 - \lambda_1'}{\lambda_2'} \times 20 \right\rfloor + 41. \tag{4.5}$$

Because $\lambda_1' \leq u_2 < \lambda_1' + \lambda_2'$ when group 2 is selected, $(u_2 - \lambda_1')/\lambda_2'$ is a uniform random variate on $[0, 1)$, and $(u_2 - \lambda_1')/\lambda_2' \times 20$ is uniformly distributed on $[0, 20)$. Therefore, $\left\lfloor \frac{u_2 - \lambda_1'}{\lambda_2'} \times 20 \right\rfloor$ yields an integer between 0 and 19 with equal probability (i.e., $= 1/20$), and Eq. (4.5) yields an integer between 41 and 60, each with probability $1/20$. Finally, if group 3 has been selected, we set

$$i = \left\lfloor \frac{u_2 - (\lambda_1' + \lambda_2')}{\lambda_3'} \times 30 \right\rfloor + 61 \tag{4.6}$$

such that i is an integer between 61 and 90, each with probability $1/30$.

Although we have considered an idealized scenario, the assumption that we can find groups of individuals sharing the event rate value is not unrealistic. In the homogeneous SIR model (i.e. where all individuals have the same susceptibility, infectiousness, and recovery rate), all the infectious individuals share the same event rate μ. Furthermore, we may be able to group the susceptible nodes according to their number of infectious neighbors and other factors. As events occur, the grouping will generally change and must thus be updated after the event. For example, if an infectious node recovers, then the group of infectious nodes, whose total event rate (for recovery) was μN_{I}, loses one member such that the total event rate is updated to $\mu(N_{\mathrm{I}} - 1)$.

4.3 Logarithmic-Time Event Selection in the Direct Method Using a Binary Tree

We now move on to a general method for speeding up the direct method. To speed up the implementation, we use a binary tree data structure to store the λ_i values. This allows us to select the reaction channel i that will produce the next event (Step 2 in Box 3) in $O(\log M)$ operations instead of $O(M)$ operations (Gibson & Bruck, 2000).[7] (See also Blue, Beichl, and Sullivan [1995] for earlier studies and Wong and Easton [1980] for the general case of sampling from urns with a general categorical probability distribution $\{\Pi_1, \ldots, \Pi_M\}$.) Because the other steps of the direct method typically have constant time complexity,

---

[7] Note that each update of a λ_j value in the binary tree also takes $O(\log M)$ time, which is slower than the original algorithm's $O(1)$ runtime. However, the overall runtime of the new algorithm is logarithmic, compared to the linear time complexity of the original one, which will make a huge difference for large systems.

(a)

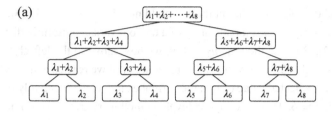

(b)

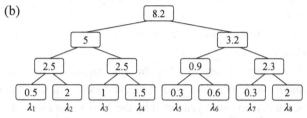

Figure 11 Binary tree for drawing i from a discrete distribution $\{\lambda_1, \ldots, \lambda_M\}$. We assume $M = 8$. (a) General case. (b) An example. The value in each nonleaf node of the tree is equal to the sum of its two child nodes' values. There are $\log_2 M + 1 = 4$ hierarchical levels.

improving the time complexity of the event selection step will speed up the entire algorithm.

The main idea is to store the λ_i values in the leaves of a binary tree and let each parent node store the sum of the values in its two child nodes (see Fig. 11(a)). By repeating this procedure for all the internal nodes of the tree, we reach the single root node on the top of the tree, to which the value $\Lambda = \sum_{i=1}^{M} \lambda_i$ is assigned. For simplicity, we assume that M is a power of 2 in Fig. 11(a) such that the tree is a perfect binary tree, that is, a binary tree where every level is completely filled. In fact, M varies in the course of a single simulation in general, but if M is not a power of 2, we can simply pad leaves of the binary tree with zeros to get a perfect tree. For example, if $M = 6$, we pad the two rightmost leaves in Fig. 11(a) with $\lambda_7 = \lambda_8 = 0$. Then, these two reaction channels are never selected for event generation. If a next event changes the number of reaction channels from $M = 6$ to $M = 7$, then we fill λ_7 by a designated positive value as well as possibly have to renew the values of some of $\lambda_1, \ldots, \lambda_6$.

To determine which event occurs, we first draw a random variate u_2 from a uniform distribution on $(0, \Lambda]$. We then start from the root node in the binary tree and look at the node's left child. If u_2 is smaller than or equal to the value stored in the left child, we move to the left child and repeat the procedure. Otherwise, we subtract the value in the left child from u_2, move to the right child, and repeat. For example, if $u_2 = 5.5$ and the binary tree is as given in Fig. 11(b), we move to the right child of the root node because $u_2 > 5$. Then, we update

u_2 by subtracting the value in the left child node: $u_2 \rightarrow 5.5 - 5 = 0.5$. Note that the new u_2 value is a realization of a random variate uniformly distributed on $(0, 3.2]$. Because the new $u_2 < 0.9$, we next move to the left child of the node with value 3.2. We repeat this procedure until we reach a leaf. This leaf's index is the selected value of i. In the current example, we eventually reach the leaf node $i = 6$. There are $\log_2 M$ levels in the binary tree if we do not count the root node of the tree. Therefore, determining a value of i given u_2 requires $O(\log M)$ time.

Once we have carried out the event associated with the selected i value, we need to update the λ_j values that are affected by the event (typically including λ_i). In the binary tree we can complete the updating locally for each j, that is, by only changing the affected leaf and its parent nodes in the tree. For example, if λ_2 changes due to an event generated by the ith reaction channel, then, first of all, we replace λ_2 by the new value. Then, we need to replace the internal node of the binary tree just above λ_2 by a new value owing to the change in the value of λ_2. For example, suppose that the new value of λ_2 is 2.4, which is 0.4 larger than the previous λ_2 value ($= 2$ as shown in Fig. 11(b)). Then, we need to increase the value of the parent of λ_2 by 0.4 so that we replace 2.5 by 2.9. We repeat this procedure up through the hierarchical levels of the tree and update the values of the relevant internal nodes and finally that of the root node. Therefore, we need to update only $\log_2 M + 1$ values per λ_j value that changes. This is a small number compared to the total number of nodes in the binary tree, which is $2M - 1$. Even if we need to update λ_j for several j's, the total number of nodes in the binary tree to be updated is typically still small compared to $2M - 1$. (However, if we need to update a large fraction of the λ_j, the number of updates may become comparable to or even surpass $2M - 1$.) The steps for implementing the direct method with binary tree search and updating of the tree are shown in Box 4.

Box 4 Gillespie's Direct Method with Binary Tree Search and Updating of the Tree

0. Initialization:
   (a) Define the initial state of the system, and set $t = 0$.
   (b) Calculate the rate λ_j for each reaction channel $j \in \{1, \ldots, M\}$.
   (c) Initialize the binary tree:
       i. Store each λ_j in a leaf of a perfect binary tree with $2^{\lceil \log_2 M \rceil}$ leafs, where $\lceil \log_2 M \rceil$ denotes the smallest integer larger than or equal to $\log_2 M$.

    ii. Fill the remaining leaf nodes with zeros.

    iii. Move up through the remaining $\lceil \log_2 M \rceil$ levels of the tree, setting the value of each node equal to the sum of the values of its two child nodes.

    iv. The value in the root node is equal to the total rate $\Lambda = \sum_{j=1}^{M} \lambda_j$.

1. Draw a random variate u_1 from a uniform distribution on $(0, 1]$, and generate the waiting time by $\tau = -\ln u_1 / \Lambda$.

2. Binary tree search:

    (a) Draw u_2 from a uniform distribution on $(0, \Lambda]$.

    (b) Start from the root node.

    (c) If $u_2 \leq a_l$, where a_l is the value in the left child of the current node, then go to the left child. Otherwise, set $u_2 \to u_2 - a_l$ and go to the right child.

    (d) Repeat Step (c) until a leaf node is reached. The index i of the leaf node gives the reaction channel that produces the next event.

3. Perform the event on reaction channel i.

4. Advance the time by setting $t \to t + \tau$.

5. Update λ_i as well as all other λ_js that are affected by the produced event.

6. Update the binary tree:

    (a) For a reaction channel whose rate λ_j changes, set $\Delta \lambda_j = \lambda_j^{(\text{new})} - \lambda_j^{(\text{old})}$, where $\lambda_j^{(\text{new})}$ and $\lambda_j^{(\text{old})}$ are the new and old event rates, respectively.

    (b) Increase the value of the jth leaf and all its parents including the root node of the tree by $\Delta \lambda_j$.

    (c) Repeat Steps (a) and (b) for all reaction channels j to be updated.

7. Return to Step 1.

4.4 Next Reaction Method: Logarithmic-Time Version of the First Reaction Method

It is also possible to make the runtime of the first reaction method scale logarithmically with the number of reaction channels, that is, to make it have a $O(\log M)$ time complexity. The improvements, collectively referred to as the *next reaction method*, were proposed in Gibson and Bruck (2000). Because both finding the smallest waiting time (Step 1 in Box 2) and updating the waiting times (Step 5 in Box 2) have $O(M)$ time complexity, we need to decrease the complexity of both steps to reduce the time complexity of the entire algorithm. The next reaction method implements three distinct improvements of the steps of the first reaction method. We describe each in turn.

4.4.1 Switch to Absolute Time

One can make Step 5 of the first reaction method (see Box 2) more efficient simply by switching from storing the putative waiting times τ_j^{put} for each reaction channel j to storing the putative absolute times of the next event, denoted by t_j^{put}. Analogous to the original first reaction method, the event with the smallest time $t = \min\{t_1^{\text{put}}, \ldots, t_M^{\text{put}}\}$ is selected to happen next, and the current time is set to t. Following the event, we only need to draw new waiting times for the reaction channel that generated the event, i, as well as for other reaction channels that are affected by the event. For these reaction channels, we reset the putative absolute time of the next event by adding the new waiting time drawn to the current time t. Thus, for each reaction channel j that must be updated, we set $t_j^{\text{put}} \rightarrow t + \tau_j^{\text{put}}$, where τ_j^{put} is the new putative waiting time drawn. There is no need to update the putative absolute times of the next event for the other reaction channels because the absolute times of the next event for these reaction channels do not depend on the current time.

In comparison, the putative waiting time until reaction channel j will generate an event, namely τ_j^{put}, does change for all the reaction channels following an event. This is because the waiting times are measured relative to the current time and thus change whenever the time advances. Therefore, the putative waiting times for all reaction channels need to be updated after each event in the original first reaction method. Because only a small number of reaction channels are affected by each event on average (except for systems that are densely connected or are in a critical phase), the use of the absolute time in place of the waiting time reduces the complexity of this step from $O(M)$ to $O(1)$.

4.4.2 Reuse Putative Times to Generate Fewer Random Numbers

Gibson and Bruck also developed a procedure for generating new putative waiting times τ_j^{put} for the reaction channels that are different from i and are affected by the event. Their idea is to reuse the old putative event time for each affected reaction channel. With this new procedure, one does not have to generate a new random variate to determine the new putative waiting time τ_j^{put} for each of these reaction channels. Let us denote by M' the number of reaction channels affected by the event in the ith reaction channel besides i itself. Then, this new procedure brings the number of random variates generated per reaction down from $M' + 1$ for the first reaction method to one for the next reaction method. While the introduction of this procedure does not change the computational complexity of the step, which remains $O(1)$, the reduction in the actual computation time may be considerable when pseudorandom number generation is much slower than arithmetic operations. However, this is generally less of a concern for newer pseudorandom number generators than it was earlier.

The procedure takes advantage of the memoryless property of Poisson processes (see Box 1). Suppose that the rate for reaction channel j has changed from $\lambda_j^{(\text{old})}$ to $\lambda_j^{(\text{new})}$ owing to the event that has occurred in reaction channel i. The memoryless property means that if the reaction channel's rate had remained unchanged, that is, if $\lambda_j^{(\text{new})} = \lambda_j^{(\text{old})}$, then the waiting time until the next event for reaction channel j, that is, $\tau_j^{\text{put}} = t_j^{\text{put}} - t$, would follow the same exponential distribution as that of the original waiting time. Furthermore, any rescaling of an exponentially distributed random variable, $\tau' = a\tau$, is also an exponentially distributed variable, with a rescaled rate $\lambda' = \lambda/a$. Thus, we define the new waiting time

$$\tau_j^{\text{put(new)}} \equiv t_j^{\text{put(new)}} - t, \tag{4.7}$$

which is related to the old waiting time by the rescaling

$$\tau_j^{\text{put(new)}} = \frac{\lambda_j^{(\text{old})} \tau_j^{\text{put(old)}}}{\lambda_j^{(\text{new})}}. \tag{4.8}$$

This $\tau^{(\text{new})}$ follows the desired distribution of the new waiting time, that is,

$$p\left(\tau^{\text{put(new)}}\right) = \lambda_j^{(\text{new})} e^{-\lambda_j^{(\text{new})} \tau_j^{\text{put(new)}}}. \tag{4.9}$$

The combination of Eqs. (4.7), (4.8), and the definition $\tau_j^{\text{put(old)}} = t_j^{\text{put(old)}} - t$ implies that, for the reaction channels j whose rates have changed, we can generate the new putative time of the next event according to

$$t_j^{\text{put(new)}} = \frac{\lambda_j^{(\text{old})} \left(t_j^{\text{put(old)}} - t \right)}{\lambda_j^{(\text{new})}} + t. \tag{4.10}$$

4.4.3 Use an Indexed Priority Queue for Selecting the Next Event

Increasing the efficiency of Step 1 of the first reaction method, which finds the reaction channel with the smallest waiting time, is more involved than the first two improvements described in Sections 4.4.1 and 4.4.2. It relies on a data structure similar to the binary tree discussed in Section 4.3. Gibson and Bruck dubbed this structure an *indexed priority queue*, which is a binary heap, that is, a type of binary tree that is optimized for implementing a priority queue, coupled to an index array (see Fig. 12). The binary heap stores the putative times of the next event, t_i^{put}, for all reaction channels, ordered from the

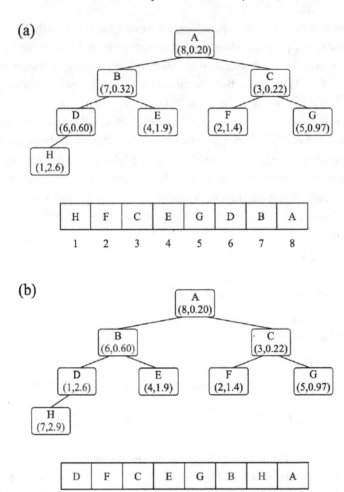

Figure 12 Indexed priority queue for storing putative reaction times in the next reaction method. (a) Example of an indexed priority queue. The indexed priority queue consists of a binary heap (top) and an index array (bottom). The binary heap contains tuples (i, t_i^{put}), where i is the reaction channel number and t_i^{put} the putative time when it would generate its next event. The nodes in the binary heap are ordered vertically by the value of t_i^{put} they store. The index array points to the node in the binary heap that corresponds to each reaction channel. (b) Configuration of the indexed priority queue after the value of t_7^{put} has been updated to 2.9 and the values stored in the nodes have been rearranged to satisfy the vertical ordering of the t_i^{put} values. All entries of the binary heap and the index array that the updating has affected are marked in red.

smallest to the largest, and provides lookup of the smallest of them, namely $\min\left\{t_1^{\text{put}},\ldots,t_M^{\text{put}}\right\}$, in $O(1)$ time. The index array contains pointers to each reaction channel's position in the binary heap to provide fast updating of the t_i values, that is, in $O(\log M)$ time.

The binary heap is a complete binary tree that stores a pair (i,t_i^{put}) in each node and is ordered such that each node has a t_i^{put} value that is smaller than that of both its children and larger than that of its parent, as shown in Fig. 12(a). Therefore, the heap stores the smallest t_i^{put} value in the root node. This implies that finding the smallest putative event time requires only a single operation, that is, it has $O(1)$ time complexity.

Because the nodes in the binary heap are not ordered by their reaction channel number i, the index array (see Fig. 12) stores for each reaction channel i a pointer to the position of the node in the binary heap that corresponds to i. Specifically, the ith entry of the index array points to the node in the binary heap that contains (i,t_i^{put}). For example, in Fig. 12(a), the reaction channel 2 is located at node F in the tree. The index array removes the need to search through the binary heap to locate a given reaction channel and the corresponding event time. The index array thus enables us to find the nodes that need to be updated after an event in $O(1)$ time.

After updating the waiting time in a given node of the binary heap, we may need to update the ordering of the nodes in the binary heap to respect the order of putative event times across the hierarchical levels (i.e., descending order as one goes from any leaf node toward the root node). We perform this reordering by "bubbling" the values up and down: starting at the node whose value has changed, corresponding to reaction channel j, say, we repeat either Step (1) or (2) in the following list, depending on the value of t_j^{put}, until one of the stopping conditions is satisfied.

(1) If the new t_j^{put} value is smaller than the t^{put} value stored in its parent node, swap (j,t_j^{put}) with the value in the parent node, and also swap the two nodes' pointers in the index array. We repeat this procedure for the parent node.

(2) Otherwise, that is, if the new t_j^{put} value is larger than or equal to the t^{put} value stored in its parent node, compare t_j^{put} to the t^{put} values in its two child nodes. If t_j^{put} is larger than the minimum of the two, swap (j,t_j^{put}) with the values in the child node that attains the minimum, and also swap the two nodes' pointers in the index array. We repeat this procedure for the child node that attained the minimum of t^{put} before the swapping.

- **Stopping conditions:** We repeat the procedure until the new t_j^{put} value is larger than its parent's and smaller than both of its children's. Alternatively,

---

Box 5 Bubbling Algorithm

bubbling(node n):

- If the t^{put} value in the node n is smaller than the t^{put} value in n's parent node, parent(n), then

    (a) swap n and parent(n), and update the index array correspondingly;

    (b) run bubbling(parent(n)).

- Else if the t^{put} value in n is larger than the smaller t^{put} value of its two children, then

    (a) swap n and the corresponding child node, min_child(n), and update the index array correspondingly;

    (b) run bubbling(min_child(n)).

- Else, stop the bubbling algorithm.

---

if the value (j, t_j^{put}) has bubbled up to the root node or down to a leaf, we also terminate the procedure.

This procedure allows us to update the binary heap in $O(\log M)$ time for each reaction channel whose event rate has changed following an event. This is the most costly part of the algorithm. So, the next reaction method improves the overall runtime of the first reaction method from $O(M)$ to $O(\log M)$. Box 5 shows an implementation of the bubbling algorithm.

To illustrate the bubbling procedure we turn to the example shown in Fig. 12(a). Suppose that the putative event time for reaction channel 7 changes from $t_7^{put} = 0.32$ to $t_7^{put} = 2.9$ following an event. We first update the value in node B of the binary heap. We then compare the value of t_7^{put} to the value in the parent node, node A, in Step (1). Because t_7^{put} is larger than the value stored in node A, we then compare t_7^{put} to the values stored in node B's two child nodes in Step (2). Because t_7^{put} is larger than the values in both the child nodes, we swap the values with the node containing the smallest of the two, which is node D, containing $t_6^{put} = 0.60$. We also update the index array by swapping the pointers of reaction channels 6 and 7. We then repeat the procedure for node D, which now contains $(7, t_7^{put})$. Because we just swapped the content of nodes B and D, we know that t_7^{put} is larger than the value stored in node D's parent node (i.e., node B). So, we compare t_7^{put} to the value stored in the only child node of node D, namely, node H. We find that t_7^{put} is larger than the value stored in node H (i.e., $t_1^{put} = 2.6$). Therefore, we swap the content of the two nodes, and we update the index array by swapping the pointers to channels 1 and 7. Because t_7^{put} is now stored in a leaf node, a stopping condition is satisfied, and we stop the procedure. Figure 12(b) shows the indexed priority queue after being updated.

One can also use the bubbling operation to initialize the indexed priority queue by successively adding nodes corresponding to each reaction channel. Because we initially need to add M values of t_i^{put} (with $i = 1,\ldots,M$), the initialization using bubbling takes $O(M\log M)$ time. More efficient methods to initialize the priority queue exist (Chen et al., 2012). However, one runs the initialization only once during a simulation. Therefore, using an efficient initialization method would typically not much contribute to the algorithm's runtime as a whole.

The binary search tree for the direct method and the binary heap for the next reaction method are both binary tree data structures and accelerate search. However, their aims are different. The binary tree for the direct method enables us to efficiently draw i with probability Π_i, and the tree holds and updates all the λ_i values ($i = 1,\ldots,M$). The binary heap used in the next reaction method enables us to efficiently determine the i that has the smallest t_i^{put} value, and the tree holds and updates all the t_i^{put} values ($i = 1,\ldots,M$). In both these structures, updating the values stored in the node (i.e., λ_j for the direct method or t_j^{put} for the next reaction method) following an event is *less* efficient than for the linear arrays used in their original implementation. Specifically, updating a value in the tree structures takes $O(\log M)$ time as opposed to $O(1)$ time for the linear array. However, the improved direct and the next reaction methods still have an overall $O(\log M)$ time complexity. In contrast, the original direct and first reaction methods have an overall $O(M)$ time complexity due to the linear search, which costs $O(M)$ time. For large systems, the time saved by getting rid of the linear search is larger than the added overhead.

Finally, note that the binary heap has only M nodes, which contrasts with the binary tree used for the direct method (Section 4.3), which has $2M-1$ nodes in the ideal case of M being a power of 2. However, because the nodes in the binary heap are ordered according to their t_i^{put} value and not their index, the binary heap also needs to store the reaction channel's index i in each node. In addition to the binary heap, the indexed priority queue also needs to store the index array, representing an additional M values. Thus, the indexed priority queue stores $3M$ values, M of which are floating-point numbers and $2M$ are integers. In contrast, the binary tree stores $2M - 1$ floating-point numbers. Therefore, the memory footprints of the indexed priority queue and the binary tree are similar.

The steps for implementing the next reaction method are shown in Box 6.

Box 6 Next Reaction Method

0. Initialization:

   (0) Define the initial state of the system, and set $t = 0$.

   (b) Calculate the rate λ_j for each reaction channel $j \in \{1, \ldots, M\}$.

   (c) Draw M random variates u_j from a uniform distribution on $(0, 1]$.

   (d) Generate a putative event time $t_j^{\text{put}} = -\ln u_j / \lambda_j$ for each j.

   (e) Initialize the indexed priority queue:

   Sequentially for each reaction channel j, add a node containing the values (j, t_j^{put}) to the binary heap and its position to the index array, by performing the bubbling algorithm (see Box 5).

1. Select the reaction channel i corresponding to the root node in the heap (which has the smallest t_i^{put}).

2. Perform the event on reaction channel i.

3. Advance the time $t \to t_i^{\text{put}}$.

4. Update λ_i and all other λ_j that are affected by the event produced.

5. Update the indexed priority queue:

   (a) Draw a new putative waiting time for reaction channel i according to $\tau_i^{\text{put}} = -\ln u / \lambda_i$, with u being drawn from uniform distribution on $(0, 1]$, and set $t_i^{\text{put}} = t + \tau_i^{\text{put}}$.

   (b) Generate new t_j^{put} values for other reaction channels j whose λ_j has changed, according to $t_j^{\text{put(new)}} \to \lambda_j^{\text{(old)}} \left(t_j^{\text{put(old)}} - t \right) / \lambda_j^{\text{(new)}} + t$.

   (c) For each reaction channel j whose t_j^{put} value has changed (including i), look up in the index array the node n that stores (j, t_j^{put}) in the binary heap; update t_j^{put} in the node n, and run the bubbling algorithm (see Box 5) to reorder the heap and update the index array.

6. Return to Step 1.

4.5 Composition and Rejection Algorithm to Draw the Next Event in the Direct Method

Let us discuss a third method to draw an event i with probability Π_i from a categorical distribution $\{\Pi_1, \ldots, \Pi_M\}$ in the direct method. The idea is to use the so-called composition and rejection (CR) algorithm (Schulze, 2008; Slepoy, Thompson, & Plimpton, 2008). This is a general method to sample a random variate that obeys a given distribution (von Neumann, 1951), which typically has a constant time complexity, namely, $O(1)$, and thus can be fast even for large systems.

The idea is to first represent the categorical distribution $\{\Pi_1, \ldots, \Pi_M\}$ as a bar graph. The bar graph corresponding to the distribution given by Fig. 8 is shown in Fig. 13. The total area of the bar graph is equal to 1. We consider a rectangle that bounds the entirety of the bar graph, shown by the dotted lines in Fig. 13. Then, we draw two random variates, denoted by u_3 and u_4, from the uniform distribution on $(0, 1]$ and consider the point $(u_3 M, u_4 \Pi_{max})$, where $\Pi_{max} = \max\{\Pi_1, \ldots, \Pi_M\}$. By construction, the point drawn is distributed uniformly (i.e., without bias) in the rectangle. If the point happens to be inside the area of the bar graph, it is in fact a uniformly random draw from the bar graph. Thus, the probability for the point to land inside the ith bar is proportional to Π_i in this case. The composition and rejection algorithm uses this property to draw the event that happens, without having to iterate over a list (or a binary tree) of λ_i values, by simply judging which bar the point falls inside. In the example shown in Fig. 13, the point drawn, shown by the filled circle, belongs to Π_2, so we conclude that reaction channel 2 has produced the event. If the point does not fall inside any bar (e.g., the triangle in Fig. 13), we then reject this point and obtain another point by redrawing two new uniform random variates. In practice, we find a putative reaction channel to produce the event by $i = \lfloor u_3 M \rfloor$ and adopt it if $\Pi_i \leq u_4 \Pi_{max}$; we reject it otherwise. The steps of the CR algorithm for general cases are shown in Box 7. These steps replace Step 2 of the direct method in Box 3. We note that the meaning of the rejection here is the same as that for the rejection sampling algorithm (see Section 2.8) but that the two algorithms are otherwise different.

If the area of the bar graph, which is always equal to 1, is close to the area of the rectangle, the CR algorithm is efficient. This is because rejection then occurs with a small probability, and we only waste a small fraction of the random variates u_3 and u_4, whose generation is typically the most costly part of the algorithm. In Fig. 13, the rectangle has area of $4 \times 0.4 = 1.6$. Therefore, one rejects $1 - 1/1.6 = 3/8$ of the generated random points on average. If the area of the box is much larger than one, which will generally happen when the event rates are heterogeneous and a few rates are much larger than the majority, then the CR algorithm is not efficient.

In Schulze (2008) and Slepoy et al. (2008), the authors went further to improve the CR algorithm to reduce the rejection probability. The idea is to organize the individual bars such that bars of similar heights are grouped together into a small number of groups and then to draw a rectangle to bound each group of bars. The probability for a group to generate the next event is proportional to the sum of the areas of the individual bars in the group. Because the number of groups is small, one can efficiently select the group that generates an event using a simple linear search. In many cases the number of groups

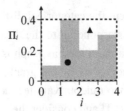

Figure 13 Bar graph for the composition and rejection algorithm. The height of each bar represents Π_i. Like in Fig. 8, we assume $N = 4$, $\Pi_1 = 0.1$, $\Pi_2 = 0.4$, $\Pi_3 = 0.2$, and $\Pi_4 = 0.3$. The two points uniformly randomly sampled from the dotted rectangle are shown by a circle and triangle.

Box 7 Composition and Rejection Algorithm

1. Generate two uniform random variates $u_3, u_4 \in (0, 1]$.
2. Set $i = \lfloor u_3 M \rfloor$.
3. If $u_4 \Pi_{\max} \leq \Pi_i$, we conclude that the ith reaction channel produces the event. Otherwise, return to Step 1.

---

does not depend on M, and this step thus has constant time complexity, $O(1)$. We then apply the original CR algorithm, given in Box 7, inside this group to select the individual reaction channel that generates the next event. This step is necessarily efficient since the reaction channels were grouped to have similar rates, so the area of the box corresponding to the group is not much larger than one. This implementation of the CR algorithm conserves its $O(1)$ time complexity. It makes the rejection step more efficient at the cost of requiring an additional random variate and having to iterate through the list of groups to select the one that generates the event.

In both the original and improved CR algorithms, the time to determine i does not depend on M, so it has $O(1)$ time complexity in terms of M. In practice, the efficiency of the algorithm depends on the probability of rejection and on the complexity of regrouping the bars in the case of the improved CR algorithm.

4.6 Recycling Pseudorandom Numbers in the Direct Method

Each iteration of the direct method requires the generation of two uniform random variates, u_1 and u_2; one for generating the waiting time, and another for selecting the reaction channel that produces an event. Generating a pseudorandom number is generally more costly than simple arithmetic operations. However, as we mentioned in Section 4.4, recent pseudorandom number generators have substantially reduced the computational cost of generating random numbers. Nevertheless, there may be situations where it is preferable to generate as few random numbers as possible. Yates and Klingbeil (2013) proposes a method to recycle a single pseudorandom number to generate both u_1 and u_2 (see also Masuda and Rocha [2018]). The method works as follows.

First, we generate a uniform random variate on $(0, \Lambda]$ denoted by u_2, where $\Lambda = \sum_{j=1}^{M} \lambda_j$. Second, we determine the reaction channel i that produces the event, which satisfies $\sum_{j=1}^{i-1} \lambda_j < u_2 \leq \sum_{j=1}^{i} \lambda_j$ (see Step 2 in Box 3). These steps are the same as those of the direct method. Now, we exploit the fact that $u_2 - \sum_{j=1}^{i-1} \lambda_j$ is uniformly distributed on $(0, \lambda_i]$ given that the reaction channel i has been selected. This is because u_2 is uniformly distributed between $\sum_{j=1}^{i-1} \lambda_j$ and $\sum_{j=1}^{i} \lambda_j$ (which is equal to $\lambda_i + \sum_{j=1}^{i-1} \lambda_j$). Therefore, we set

$$u_1 = \frac{u_2 - \sum_{j=1}^{i-1} \lambda_j}{\lambda_i}, \tag{4.11}$$

which is a uniform random variate on $(0, 1]$. Then, we generate the waiting time by $\tau = -\ln u_1 / \Lambda$ (see Step 1 in Box 3).

There are two remarks. First, we need to determine i and then determine τ with this method. In contrast, one can first determine either i or τ as one likes in the original direct method. Second, by generating two pseudorandom numbers from a one pseudorandom number, one is trading speed for accuracy. The variable u_1 has a smaller number of significant digits (i.e., less accuracy) compared to when one generates u_1 directly using a pseudorandom number generator as in the original direct method. However, this omission probably does not cause serious problems in typical cases as long as M is not extremely large because the u_1 generated by the recycling direct method and the original direct method differ only slightly in the numerical value.

4.7 Network Considerations

In networks, where different nodes may have different degrees, and the number of infectious neighbors may be different even for same-degree nodes, it may be difficult to bookkeep, select, and update the λ_j values efficiently. So, we need careful consideration of these steps when simulating stochastic processes in networks (Kiss et al., 2017a; St-Onge et al., 2019). A particular problem that arises in dense networks or for dynamic processes in heterogeneous networks that are close to a critical point is that the average number of reaction channels that are affected by each event can become extremely large. Concretely, it may be proportional to the number of reaction channels, M. In this case updating the event rates λ_j (Step 4 in Boxes 2 and 6 and Step 5 in Boxes 3 and 4) no longer has an $O(1)$ time complexity but $O(M)$. Because none of the methods just discussed improves this step, they then cannot improve the time complexity of the classic Gillespie algorithms in this situation.

Carefully designed event-based simulations, which are similar in spirit to the first reaction method, can significantly accelerate exact simulations of coupled jump processes, even for dense or heterogeneous networks and including the case of coupled non-Poissonian renewal processes (see also Section 5.5). An important assumption underlying this approach is that, once a node is infected, it will recover at a time that is drawn according to the distribution of recovery times regardless of what is going to happen elsewhere in the network. In this way, one can generate and store the recovery time of this node in a priority queue to be retrieved when the time comes. Codes for the SIR model and the susceptible-infectious-susceptible (SIS) model (i.e., individuals may get reinfected after a recovery) as well as for generating animation and snapshot figures are available at Kiss, Miller, and Simon (2017b). The corresponding pseudo-code and explanation are available in Appendix A of Kiss, Miller, and Simon (2017a).

Another idea that can be used for speeding up simulations in this case is that of phantom processes, which is to assign a positive event rate to types of events that actually cannot occur. For example, an infectious node attempts to infect an already infectious or recovered node. If such an event is selected in a single step of the Gillespie algorithm, nothing actually occurs, and so the event is wasted. However, by designing such phantom processes carefully, one can simplify the updating of the list of all possible events upon the occurrence of events, leading to overall saving of computation time (Cota & Ferreira, 2017).

St-Onge and colleagues have advanced related simulation methods in three main aspects (St-Onge et al., 2019). First, they noted the fact that, in the case of the SIR and SIS models, any event, namely either the infection or the recovery event, involves an infectious individual. Therefore, we can reorganize the set of possible events such that they are grouped according to the individual infectious nodes. In other words, an infectious node either recovers with rate μ or infects one of its susceptible neighbors with rate β. Therefore, the total event rate associated to an infectious node v_i is equal to $k_{i,S}\beta + \mu$, where $k_{i,S}$ is the number of susceptible neighbors of v_i. In this manner, one only has to monitor N_I reaction channels during the process. Their second idea is to use the CR algorithm (Section 4.5). We have

$$\Pi_i = \frac{k_{i,S}\beta + \mu}{\sum_{j;v_j \text{ is infected}} (k_{j,S}\beta + \mu)} \tag{4.12}$$

for infectious nodes v_i. Because infected nodes tend to be large-degree nodes (Barrat et al., 2008; Pastor-Satorras et al., 2015; Pastor-Satorras & Vespignani,

2001), some Π_i's tend to be much larger than other Π_i's. To accelerate the sampling of the event i that occurs in each step in this situation, they employed the improved CR algorithm of Slepoy and colleagues (2008). Third, they employed phantom processes, corresponding to infections of already infectious nodes.

Their algorithm has a time complexity of $O(\log \log M)$ and is thus efficient in many cases. Their code, whose computational part is implemented in C++ for efficiency and whose interface is in Python, is available on Github (St-Onge, 2019).

4.8 Tau-Leaping Method

There are various other algorithms that are related to the Gillespie algorithms and introduce some approximations to speed up the simulations. We briefly review just one such method here, the *tau-leaping method*, which Gillespie proposed in 2001. The method works by discretizing time into intervals of some chosen length, Δt. In a given interval, the method draws a random variate to determine how many events have happened for each process i, denoted by $\bar{n}_i$, and then updates the state of the system (Gillespie, 2001). For example, in a chemical reaction system, $\bar{n}_i$ is the increment in the number of molecules of the ith species. Under the assumption that each λ_i stays constant between $[t, t + \Delta t]$, where t is the current time, $\bar{n}_i$ obeys a Poisson distribution with mean $\lambda_i \Delta t$. In other words, the number is equal to $\bar{n}_i$ with probability $(\lambda_i \Delta t)^{\bar{n}_i} e^{-\lambda_i \Delta t} / \bar{n}_i!$. It is desirable to make Δt large enough to reduce the computation time as much as possible. On the other hand, Δt should be small enough to guarantee that λ_i stays approximately constant in each time window of length Δt in order to assure the accuracy of the simulation. In simulations of social dynamics, the tap-leaping methods are probably not relevant in most cases because a single event on the ith reaction channel typically produces the state change of, for example, the ith individual. Then, one needs to renew the λ_i value. If this is the case, we cannot use the same λ_i to produce multiple events on the ith reaction channel.

For other methods to accelerate the Gillespie algorithm or related algorithms, we refer to the review paper by Goutsias and Jenkinson (2013).

4.9 Pseudorandom Number Generation

All stochastic simulation algorithms including the Gillespie algorithms rely on the generation of random numbers. We here give a brief and practical introduction to computer generation of random numbers for application to Monte Carlo simulations. We will not address the technical workings of random number generators here. An authoritative introduction to the subject is found in Chapter 7 of Press and colleagues (2007).

A pseudorandom number generator (PRNG) is a deterministic computer algorithm that generates a sequence of approximately random numbers. Such numbers are not truly random. So, we refer to them as pseudorandom (i.e., seemingly random) to distinguish them from numbers generated by a truly random physical process. However, the numbers need not be truly random in most applications. They just need to be random enough. In the context of Monte Carlo simulations, a working criterion for what constitutes a good PRNG is that simulation results based on it are indistinguishable from those obtained with a truly random source (Jones, 2010).[8] Not all PRNGs satisfy this requirement. In fact, many standard PRNG algorithms still in use today have been shown to have serious flaws. To make this volume self-contained, we here explain some simple good practices to be followed and pitfalls to be avoided to ensure that the pseudorandom numbers that our simulations rely on are of sufficient quality. For a more detailed, yet not too technical, introduction to good and bad practices in pseudorandom number generation, we recommend Jones (2010).

What constitutes a good PRNG depends on the application. For example, good PRNGs for Monte Carlo simulations are generally not random enough for cryptography applications. Conversely, while cryptographically secure PRNGs produce high-quality random sequences, they are generally much slower and are thus not optimal for Monte Carlo simulations. As this example suggests, the choice of PRNG involves a trade-off between the statistical quality of the generated sequence and the speed of generation. Nevertheless, many PRNGs now exist that are both fast and produce sequences that are sufficiently random for any Monte Carlo simulation. Ensuring that you use a good PRNG essentially boils down to checking two simple points: (1) do not use your programming language's standard PRNG, and (2) properly seed the PRNG.

First, the most important rule to follow when choosing a PRNG is to never use your programming language's standard PRNG! To ensure backwards compatibility, the standard random number generators in many programming languages are based on historical algorithms that do not produce good pseudorandom number sequences. (A nonexhaustive list of languages with bad PRNGs is found in Jones [2010].) Extensive test suites such as TestU01 (L'Ecuyer & Simard, 2007) and Dieharder (Brown, Eddelbuettel, & Bauer, 2021) have been developed for testing the statistical quality of PRNGs.

Several fast and high-quality PRNGs are implemented in standard scientific computing libraries. So, using a good PRNG is as simple as importing it from

---

[8] An entirely equivalent definition is that any two (good) PRNGs should lead to statistically the same results as the simulations (Press et al., 2007).

one of these libraries.[9] For example, we obtained all simulations performed in Python that are shown in this tutorial using the 64-bit Permuted Congruential Generator (PCG64) (O'Neill, 2014), which is available in the standard NumPy library. For C++ code (see Section 4.10), we used the Mersenne Twister 19937 algorithm with improved initialization,[10] namely mt19937ar (Matsumoto, 2021; Matsumoto & Nishimura, 1998), which is available as part of the Boost and Libstdc++ libraries.

Second, we should properly seed the PRNG. PRNGs rely on an internal state, which they use to generate the next output in the random number sequence. The internal state is updated at each step of the algorithm. At the first use of the PRNG, one must initialize, or *seed*, the internal state. Proper initialization is crucial for the performance of a PRNG. In particular, earlier PRNGs suffered from high sensitivity to the seed value. For example, the original implementation of the Mersenne Twister algorithm is hard to seed due to its slow mixing time. This means that, if the bitstring corresponding to the initial state is not random enough (e.g., if it contains mostly zeros), up to the first one million generated numbers can be nonrandom. Recently proposed PRNGs generally do not show the same pathologies. So, as long as one uses a recently proposed, good PRNG, seeding the generator is not a problem except when one runs simulations in parallel. The initial seeding problems of the Mersenne Twister have been fixed in the mt19937ar version in 2002 (Matsumoto, 2021). As another example, the PCG64 PRNG is easy to seed.

We need to be more careful on how to seed each instance of the PRNG when running simulations in parallel. An often used method to select the seed for a PRNG is to generate it automatically based on the system clock. However, this is a bad idea for launching many (e.g., thousands or more) simulations

---

[9] Many sources advise to implement the PNRG for oneself in one's code. While this is instructive, we do not believe that it is necessary for starting your work with Monte Carlo simulations.

[10] The Mersenne Twister 19937 algorithm (MT19937) has long been the reference for Monte Carlo simulations. (It has now been surpassed by the last generation of PRNGs, both in terms of speed and memory requirement, and in terms of its performance on statistical randomness tests.) MT19937 passes almost all tests in the test suites (e.g., Dieharder and TestU01), but it fails a few of the more rigorous tests of randomness which recent PRNGs, such as PCG64, pass. However, these are very rigorous tests, and this nonrandomness is unlikely to pose any problem in Monte Carlo simulations (Jones, 2010). While MT19937 is a bit slower than PCG64, the speed difference is generally too small to matter much in practice. (Tests show that PCG64 is around twice as fast as MT19937 when generating 64-bit random numbers (O'Neill, 2014).) Finally, MT19937 has a much larger memory footprint than other PRNGs; it keeps a 20 032-bit internal state compared to PCG64's 128-bit state, for example. This could pose a problem when running massively parallel simulations, for example, on a graphics processing unit (GPU) with a relatively small amount of memory. However, it will not be a problem when running a single or a few parallel simulations on a desktop computer.

in parallel because many simulations will then tend to be initialized with the same or nearly the same seed. Any simulations launched with the same seed will produce exactly the same results, and those launched with close seeds may produce correlated results depending on the mixing properties of the PRNG. In both cases, this is wasteful. What is worse, if we are not aware that the simulation results are correlated, we will overestimate the precision of the obtained results. The safest way to seed the PRNG for parallel simulation is to use a *jump-ahead* operation that allows us to advance the internal state of the PRNG by an arbitrary amount of steps. With this method, one can initialize the PRNG of individual simulations with states that are sufficiently far from each other such that the pseudorandom number sequences generated by the different simulations are dissimilar. Methods for parallel seeding exist for both the Mersenne Twister (Haramoto et al., 2008) and PCG64 (O'Neill, 2014). If one wants to use a PRNG for which no efficient jump-ahead method exists, a better source of randomness than the system time should be used. On Unix machines, /dev/urandom is a choice. Note, however, that all PRNGs have a fixed cycle length, after which it will repeat itself deterministically. Therefore, one should use a PRNG with a sufficiently large cycle length (at least 2^{64}) and a seed with a sufficient number of bits (at least 64) to avoid overlap between the pseudorandom number sequences generated by the PRNG.

4.10 Codes

Here we showcase some example simulations of event-based stochastic processes using the Gillespie algorithms. In practice, in scientific research in which the Gillespie algorithms are used, we often need to exactly run coupled jump processes on a large scale. For example, you may need to simulate a system composed of many agents, or you may have to repeat the same set of simulations for various parameter values to investigate the dependence of the results on the parameter values of your model. In such a situation, we often want to implement the Gillespie algorithms in a program language faster than Python. Therefore, we implemented the SIR model and three other dynamics in C/C++, which is typically much faster than in Python. We use the Mersenne Twister as the PRNG and only implement the direct method. Our C/C++ codes and the list of edges of the networks used in our demonstrations are available at Github (https://github.com/naokimas/gillespie-tutorial).

4.10.1 SIR Model

We provide codes for simulating the SIR model in well-mixed populations (sir-wellmixed.cc), for general networks using Gillespie's original

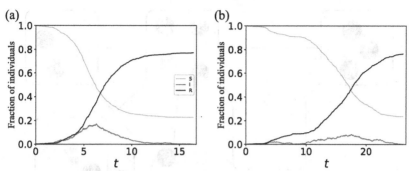

Figure 14 Time courses of the fraction of the susceptible, infectious, and recovered nodes obtained from two runs of the SIR model. We set $\beta = 0.6$ and $\mu = 3$. We used a regular random graph with $N = 1000$ nodes and the nodes' degree equal to five. In other words, each node has degree 5, and apart from that, we connect the nodes uniformly at random according to the configuration model (Fosdick et al., 2018). The network is the same for the two runs shown in (a) and (b). Each run started from the same initial condition in which a particular node was infectious and the other $N - 1$ nodes were susceptible.

direct method (`sir-net.cc`), and for general networks using the binary search tree (see Section 4.3) to speed up the selection of the events (`sir-net-binary-tree.cc`). Time courses of the fractions of the susceptible, infectious, and recovered nodes from two runs of the SIR model with $\beta = 0.6$ and $\mu = 3$ on a regular random graph with $N = 1000$ nodes are shown in Fig. 14. We started both runs from the same initial condition in which just one node, which was the same node in both runs, was infectious and the other $N - 1$ nodes were susceptible. The figure illustrates the variability of the results due to stochasticity, which is lacking in the ODE version of the SIR model (Section 2.7).

4.10.2 Metapopulation Model with SIR Epidemic Dynamics

Another example system that the Gillespie algorithms can be used for is the SIR model in a metapopulation network. Mobility may induce different contact patterns at different times. For example, we typically contact family members in the morning and evening, while we may contact workmates or schoolmates in the day time. The metapopulation model provides a succinct way to model network changes induced by mobility (Anderson & May, 1991; Colizza et al., 2006; Colizza, Pastor-Satorras, Vespignani, 2007; Diekmann & Heesterbeek, 2000; Hanski, 1998; Hufnagel, Brockmann, & Geisel, 2004). We consider a network, where a node is a patch, also called a subpopulation, which is a container of individuals, modeling for example, a home, a workplace, a sports team

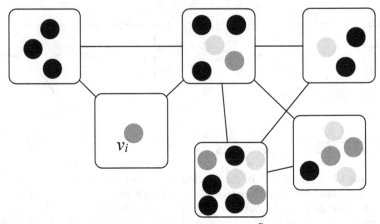

Figure 15 A metapopulation model network with $\tilde{N} = 6$ patches and $N = 25$ individuals. As in the previous similar figures, the blue, red, and brown circles represent susceptible, infectious, and recovered individuals, respectively.

meeting, or a pub. A network of patches is distinct from a network in which a node is an individual. In Fig. 15, there are $\tilde{N} = 6$ patches connected as a network. Each individual is in either the S, I, or R state and is assumed to be situated in one patch; there are $N = 25$ individuals in Fig. 15. An infectious individual infects each susceptible individual in the same patch with rate β. Crucially, an infectious individual does not infect susceptible individuals in other patches. An infectious individual recovers with rate μ regardless of who are in the same patch.

In addition, the individuals move from a patch to another. There are various mobility rules used in the metapopulation model (Masuda & Lambiotte, 2020), but a simple one is the so-called continuous-time random walk. In its simplest variant, each individual moves with constant rate D, which is often called the diffusion rate. In other words, each individual stays in the currently visited patch for a *sojourn time* τ, which follows the exponential distribution, $\psi_{\text{stay}}(\tau) = De^{-D\tau}$, before it moves to a neighboring patch. When the individual moves, it selects each neighboring patch with equal probability. For example, the infectious individual v_i in Fig. 15 moves to either of the neighboring patches with probability $1/2$ when it moves. The movements of different individuals are independent of each other, and the moving events occur independently of the infection or recovery events. Because we assumed that the time to the next move of each individual obeys an exponential distribution, we can use the Gillespie algorithms to simulate the SIR plus mobility dynamics as described by the standard metapopulation model.

We provide a code (`sir-metapop.cc`) for simulating the SIR model in the metapopulation model to which one can feed an arbitrary network structure.

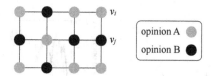

Figure 16 Schematic of the voter model on a network.

Time courses of the numbers of S, I, and R individuals are qualitatively similar to those for the standard SIR model in well-mixed populations and networks.

4.10.3 Voter Model

Another typical example of collective dynamics is the voter model (Barrat et al., 2008; Castellano, Fortunato, & Loreto, 2009; Holley & Liggett, 1975; Krapivsky, Redner, & Ben-Naim, 2010; Liggett, 1999). Suppose again that the individuals are nodes of a network. Each individual is a voter and takes either of the two states A and B, referred to as opinions (see Fig. 16). If two individuals adjacent on the network have the opposite opinions, the A individual, denoted by v_i, tries to convince the B individual, denoted by v_j, into supporting opinion A, in the same manner as an infectious individual infects a susceptible individual in the SIR model. This event occurs with rate $\beta_{B\rightarrow A}$. At the same time, v_j tries to convince v_i, who currently supports opinion A, into supporting opinion B, which occurs with rate $\beta_{A\rightarrow B}$. Clearly, the two opinions compete with each other. The time before v_j flips its opinion from B to A due to v_i obeys an exponential distribution given by $\psi_{B\rightarrow A}(\tau_{B\rightarrow A}) = \beta_{B\rightarrow A}e^{-\beta_{B\rightarrow A}\tau_{B\rightarrow A}}$. Likewise, the time before v_i flips its opinion from A to B due to v_j obeys an exponential distribution given by $\psi_{A\rightarrow B}(\tau_{A\rightarrow B}) = \beta_{A\rightarrow B}e^{-\beta_{A\rightarrow B}\tau_{A\rightarrow B}}$. If $\tau_{B\rightarrow A} < \tau_{A\rightarrow B}$ and nothing else occurs on the network for time $\tau_{B\rightarrow A}$ from now, v_j flips its opinion from B to A. This implies that v_j loses the chance to convince v_i to take opinion B because v_j itself now supports opinion A.

Such a competition occurs on every edge of the network that connects two nodes with the opposite opinions. The dynamics stop when the unanimity of opinion A or that of opinion B has been reached. Then, there is no opinion conflict in the entire population. Note that opinion B does not emerge if everybody in the network has opinion A, and vice versa. For this and other reasons, the voter model is not a very realistic model of voting or collective opinion formation. However, the model has been extensively studied since its inception in the 1970s. The most usual setting is to assume $\beta_{A\rightarrow B} = \beta_{B\rightarrow A}$ (i.e., both opinions are equally influential) and ask questions such as the time until consensus (i.e.,

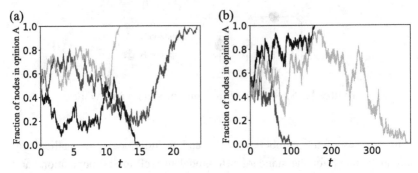

Figure 17 Time courses of the fraction of the nodes with opinion A in the voter model in three different simulations. We use a regular random graph with nodes' degree equal to five. We set (a) $N = 100$ and (b) $N = 1000$. We set $\beta_{A \to B} = \beta_{B \to A} = 1$. Each run started from the initial condition in which half the nodes were in opinion A and the other half were in opinion B. The results for three runs are shown in different colors in each panel.

unanimity) and which opinion is likely to win depending on the initial conditions. When $\beta_{A \to B} \neq \beta_{B \to A}$, the model is called the biased voter model, and an additional question to be asked is which opinion is likely to win depending on the imbalance between $\beta_{A \to B}$ and $\beta_{B \to A}$. Because there are only two types of events, associated with $\beta_{A \to B}$ and $\beta_{B \to A}$, and they occur with exponentially distributed waiting times, one can simulate the voter models, including biased ones, using the standard Gillespie algorithms.

We provide codes for simulating the voter model in well-mixed populations (voter-wellmixed.cc), for general networks using Gillespie's original direct method (voter-net.cc), and for general networks using a binary search tree (voter-net-binary-tree.cc). Time courses of the fraction of the nodes in opinion A from three runs of the unbiased voter model on a regular random graph with $N = 100$ and $N = 1000$ nodes are shown in Fig. 17(a) and Fig. 17(b), respectively. All the runs for each network started from the same initial condition in which half the nodes are in opinion A and the other half in opinion B. The figure indicates that some runs terminate with the consensus of opinion A and the others with the consensus of opinion B. It takes much longer time before a consensus is reached with $N = 1000$ (Fig. 17(b)) than with $N = 100$ (Fig. 17(a)), which is expected. Results for well-mixed populations (which one can produce with voter-wellmixed.cc) are similar to those shown in Fig. 17.

4.10.4 Lotka–Volterra Model

The Lotka–Volterra model describes dynamics of the numbers of prey and of predators under predator–prey interaction. It is common to formulate and

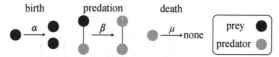

Figure 18 Rules of the stochastic Lotla–Volterra model with one prey species and one predator species.

analyze this dynamics as a system of ODEs, where the dependent variables represent the numbers of the prey and predators, and the independent variable is time. The ODE approach to the Lotka–Volterra model and its variants have been particularly useful in revealing mathematical underpinnings of oscillatory time courses of the numbers of prey and predators (Hofbauer & Sigmund, 1988; Murray, 2002). However, it is indispensable to consider stochastic versions of the Lotka–Volterra models (Dobrinevski & Frey, 2012; Gokhale et al., 2013; Parker & Kamenev, 2009) when the number of prey or of predators is small. (See Section 2.7 for a general discussion of the problems with ODE models.)

Consider a system composed of a single species of prey (which we call rabbits) and a single species of predator (which we call foxes). We denote the number of rabbits and that of foxes by N_{rab} and N_{fox}, respectively. The rules of how N_{rab} and N_{fox} change stochastically are shown schematically in Fig. 18. A rabbit gives birth to another rabbit with rate α. A fox dies with rate μ. A fox consumes a rabbit with rate β, which by definition results in an increment of N_{fox} by one. This assumption is probably unrealistic because a fox would not give birth to its cub only by consuming one rabbit. (A fox probably has to eat many rabbits to be able to bear a cub.) The model furthermore ignores natural deaths of the rabbits. These omissions are for simplicity. Because the three types of events occur as Poisson processes with their respective rates and we also assume that different types of events occur independently of each other, one can simulate the stochastic Lotka–Volterra dynamics using the Gillespie algorithms.

The extension of the Lotka–Volterra system to the case of many species is straightforward. In this scenario, a species i may act as prey toward some species and as predator toward some other species. A version of the Lotka–Volterra model for more than two species can be described by the birth rate of species i, denoted by α_i (with which one individual of species i bears another individual of the same species); the natural death rate of species i, denoted by μ_i; and the rate of consumption of individuals of species j by one individual of species i, denoted by β_{ij} (i.e., an individual of species i consumes an individual of species j with rate β_{ij}).

(a)   (b)

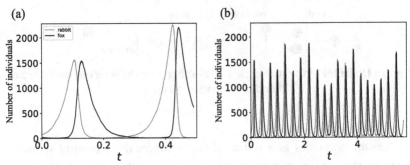

Figure 19 Number of rabbits and foxes for two different runs of the stochastic Lotka–Volterra model in a well-mixed population. In both (a) and (b), we set $\alpha = 30$, $\beta = 0.1$, $\mu = 30$, and there are initially $N_{\text{rab}} = 80$ rabbits and $N_{\text{fox}} = 20$ foxes.

We provide a code ($\texttt{lotka-volterra-wellmixed.cc}$) for simulating the stochastic single-prey single-predator Lotka–Volterra dynamics in a well-mixed population. Two sample time courses of the number of rabbits and that of foxes are shown in Fig. 19. In both runs, the initial condition was the same (i.e., $N_{\text{rab}} = 80$ rabbits and $N_{\text{fox}} = 20$ foxes). We see oscillatory behavior of both species with time lags, which is well known to appear in the Lotka–Volterra model. In Fig. 19(a), the simulation terminated when the rabbits went extinct after two cycles of wax and wane. By contrast, in Fig. 19(b), the simulation terminated when the foxes went extinct after many cycles of wax and wane. The apparent randomness in the sequence of the height of the peaks in Fig. 19(b) is due to the stochasticity of the model.

The results shown in Fig. 19 are in stark contrast with those that the ODE version of the Lotka–Volterra model would produce in two aspects. First, the two time courses from the present stochastic simulations look very different from each other due to the stochasticity of the model. The ODE version will produce the same result every time if the simulation starts from the same initial conditions and one can safely ignore rounding errors. Second, the ODE version does not predict the extinction of one species; N_{rab} or N_{fox} can become tiny in the course of the dynamics, but it never hits zero in finite time. By contrast, the stochastic-process version always ends up in extinction of either species, although it may take long time before the extinction occurs. Once rabbits go extinct, the foxes will necessarily go extinct because there is no prey for the foxes to consume. With our code, a run terminates once rabbits go extinct in this case. On the contrary, if foxes go extinct first, then the number of rabbits will grow indefinitely because the predators are gone. In either case, there is no room for foxes to survive.

5 Gillespie Algorithms for Temporal Networks and Non-Poissonian Jump Processes

Until now we have assumed that all events occur according to Poisson processes and that the interaction network, including the case of the well-mixed population, stays the same over the duration of the simulation. However, both of these assumptions are often violated in empirical social systems. In this section, we present algorithms that relax these assumptions and allow us to simulate processes with non-Poissonian dynamics and on networks whose structure evolves over time.

5.1 Temporal Networks

In general, interactions between individuals in a social system are not continually active, so the networks they define vary in time (Fig. 20). The statistics of both the dynamics of empirical temporal networks and the dynamic processes taking place on them are often strongly non-Poissonian, displaying both nonexponential waiting times and temporal correlations. Both the dynamics of empirical networks and of processes taking place on dynamic networks have been studied under the umbrella term of temporal networks (Holme, 2015; Holme & Saramäki, 2012, 2013, 2019; Masuda & Lambiotte, 2020). In the following subsections, we present several recent extensions of the direct method to temporal network scenarios. Before that, let us clarify which situations we want to extend it to.

First, empirical sequences of discrete events tend to strongly deviate from Poisson processes. In a Poisson process, the distribution of interevent times is an exponential distribution. By contrast, events in empirical human activity data often do not exhibit exponential distributions. Figure 21 shows the distribution of interevent times τ between face-to-face encounters for an individual in a primary school. For reference, we also show an exponential distribution whose mean is the same as that of the empirical data. The empirical and exponential distributions do not resemble each other. In particular, the empirical distribution is much more skewed than the exponential distribution. It thus has

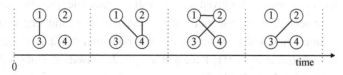

Figure 20 Schematic of a "switching" temporal network with $N = 4$ nodes. The network switches from one static graph to another at discrete points in time.

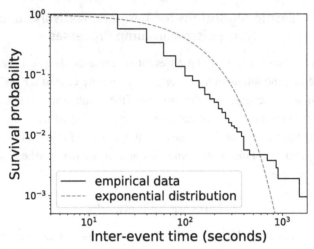

Figure 21 Survival probability of interevent times between face-to-face encounters. Solid black line: empirical data; dashed blue line: exponential distribution having the same mean as that of the empirical data. The empirical data come from the "Primary School" data set from the SocioPatterns project (Isella et al., 2011). Events are face-to-face proximity relationships between an individual and other individuals in the school. We show the survival probability, that is, $\Psi(\tau) = \int_{\tau}^{\infty} \psi(\tau')d\tau'$, instead of the distribution of interevent times, $\psi(\tau)$, because $\Psi(\tau)$ is more robust to noise in data. In other words, the vertical axis represents the fraction of the interevent times that are larger than the value specified on the horizontal axis. We selected the individual with the largest number of events and calculated all the interevent times for the selected individual. We omitted the largest interevent time, which is more than 10 times larger than the second largest one. The survival probability of the exponential distribution is given by

$$\Psi(\tau) = \int_{\tau}^{\infty} \lambda e^{-\lambda \tau'} d\tau' = e^{-\lambda \tau}.$$

a much larger chance of producing extreme values of τ, both small and large. Typically, the right tail of the distribution (i.e., at large values of τ) is roughly approximated by a power-law distribution $\psi(\tau) \propto \tau^{-\alpha}$, where $\propto$ means proportional to, and α is a constant, typically between 1 and 3. If one replaces the exponential $\psi(\tau)$, which the Gillespie algorithm and the original stochastic multiagent models assume, by a power-law $\psi(\tau)$, the results may considerably change. For example, for given β and μ values, epidemic spreading may be less likely to occur in the SIR model with interevent times τ following a power-law distribution than with ones following an exponential distribution with the same mean (Karsai et al., 2011; Kivelä et al., 2012; Masuda & Holme, 2013; Miritello, Moro, & Lara, 2011). Therefore, we are interested in simulating stochastic dynamics where the waiting times obey distributions that may differ from the exponential distribution. Such processes are called *renewal processes*.

A Poisson process is the simplest example of a renewal process. It generates events at a constant rate irrespectively of the history of the events in the past and is thus called memoryless, which is also referred to as the process being *Markovian*. General renewal processes are not memoryless and are often referred to as being non-Markovian, especially in the physics literature. Note that, even in general renewal processes, each waiting time is independent of the past ones. However, the expected waiting time until the next event depends on the time elapsed since the last event.[11] In Sections 5.2 and 5.3, we will present two algorithms that simulate stochastic dynamics when $\psi(\tau)$ can be nonexponential distributions.

Second, we may be interested in simulating a dynamic process on an empirically recorded temporal network (e.g., epidemic spread over a mobility network). We will here consider a representation of temporal networks in which the network changes discontinuously in discrete time points (Fig. 20), which we call *switching networks*. Such a representation is often practical since empirical temporal networks are generally recorded with finite time resolution and thus change only in discrete points in time. The Gillespie algorithms do not directly apply in this second case either. This is because, in switching networks, which events can occur and the rates at which they occur depend on time. In contrast, the classic Gillespie algorithms assume that the event rates stay constant in between events. We will present a temporal version of the direct method that can treat switching networks in Section 5.4.

Both non-Poissonian statistics of event times and temporally changing event rates can also occur in chemical reaction systems, for which the Gillespie algorithms were originally proposed. Several extensions have been developed in the chemical physics and computational biology literature to deal with these issues. Different from social systems, temporally changing event rates are often externally driven in such systems. For example, in cellular reaction systems the cell's volume may change over time owing to cell growth. Such a volume change leads to changes in molecular concentrations and thus to temporally evolving reaction rates, similar to temporal networks. Extensions of the Gillespie algorithms have enabled, for example, simulating chemical reaction systems with time-varying volumes (Carletti & Filisetti, 2012; Kierzek, 2002; Lu et al., 2004). There are also Gillespie algorithms for more generally fluctuating event rates (Anderson, 2007). Another common phenomenon in chemical reaction systems is delays due to, for example, diffusion-limited reactions. Such delays

---

[11] For this reason, renewal processes are formally defined as a type of semi-Markov process. We will not delve further into this distinction here. We will simply refer to processes that do not have exponentially distributed interevent times as non-Poissonian.

lead to nonexponential waiting times, and several approaches have been developed to deal with this case (Anderson, 2007; Barrio et al., 2006; Bratsun et al., 2005; Cai, 2007). While these issues are similar to those encountered in temporal networks, each has its particularities. Delays in chemical reaction systems lead to distributions of interevent times that are less skewed than the exponential distribution. In contrast, typical distributions of interevent times in social networks are *more* skewed than exponential distributions. Another difference is that external dynamics influencing chemical reaction systems are typically much slower than the reaction dynamics, while social network dynamics typically occur on the same scale or faster than the dynamics we want to simulate on networks. These facts pose specific challenges for the simulation algorithms. In fact, although some extensions of the Gillespie algorithms developed for chemical reaction systems may also be suitable for simulating multiagent systems and temporal networks, algorithms focusing specifically on the temporal network setting have emerged. We review them in this section.

5.2 Non-Markovian Gillespie Algorithm

The *non-Markovian Gillespie algorithm* is an extension of the direct method to the case in which interevent times are not distributed according to exponential distributions (Boguñá et al., 2014). It relaxes the assumption that the individual jump processes are Poisson processes and enables us to simulate general renewal processes.

We denote by ψ_i the distribution of interevent times for the ith reaction channel (where $i = 1, \ldots, M$), which we assume is a renewal process. If ψ_i is an exponential distribution, the ith renewal process is a Poisson process. If all ψ_is are exponential distributions, we can use the original Gillespie algorithms. When ψ_i is not an exponential distribution, we need to know the time $\tilde{t}_i$ since the last event for the process i to be able to generate the time to the next event for that process.

By definition, the interevent time τ_i between two successive events produced by the ith process is given by $\psi_i(\tau_i)$. We want to know the next event time, $t_i^{\text{last}} + \tau_i$, where t_i^{last} is the time of the last event on the ith reaction channel. The calculation of $t_i^{\text{last}} + \tau_i$ is not straightforward to implement using the direct method because knowing the function ψ_i for each reaction channel is not enough on its own to simulate coupled renewal processes. In fact, we must be able to calculate not only the waiting time since the last event of the ith process, but since an arbitrary time t at which another process may have generated an event. Suppose that i has not produced an event for a time $\tilde{t}_i$ after its last event. (The current time is thus $t_i^{\text{last}} + \tilde{t}_i$.) We denote by $\tilde{\tau}_i$ the waiting time until the next

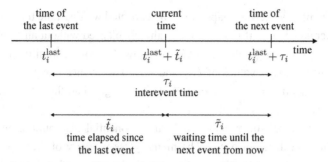

Figure 22 Schematic definition of the different notions of times employed in this section and the relations between them.

event starting from time $t_i^{last} + \tilde{t}_i$. See Fig. 22 for a schematic definition of the different notions of times. The waiting time $\tilde{\tau}_i$ does not obey $\psi_i(\tau_i)$. Instead, $\tilde{\tau}_i$ obeys the following conditional probability density with which the next event occurs at time $t_i^{last} + \tilde{t}_i + \tilde{\tau}_i$ given that no event has occurred between t_i^{last} and $t_i^{last} + \tilde{t}_i$:

$$\psi_i^{W}(\tilde{\tau}_i|\tilde{t}_i) = \frac{(\text{Probability that the next event occurs at } t_i^{last} + \tilde{t}_i + \tilde{\tau}_i)}{(\text{Probability of no event between } t_i \text{ and } t_i^{last} + \tilde{t}_i)}$$

$$= \frac{\psi_i(\tilde{t}_i + \tilde{\tau}_i)}{\Psi_i(\tilde{t}_i)}. \tag{5.1}$$

Here

$$\Psi_i(\tilde{t}_i) = \int_{\tilde{t}_i}^{\infty} \psi_i(\tau')d\tau' \tag{5.2}$$

is the survival probability, that is, the probability that the interevent time is larger than $\tilde{t}_i$. The preceding argument shows that the waiting time to the next event for each process explicitly depends on $\tilde{t}_i$. Therefore, we need to record when the last event has happened (i.e, t_i^{last}, which is $\tilde{t}_i$ before the current time) to generate the waiting time.

As an example, we consider a power-law distribution of interevent times given by

$$\psi_i(\tau_i) = \frac{\alpha - 1}{(1 + \tau_i)^{\alpha}}. \tag{5.3}$$

By substituting Eq. (5.3) into Eq. (5.1), we find the probability distribution for the waiting time $\tilde{\tau}_i$ until the ith renewal process generates its next event given that a time $\tilde{t}_i$ has already elapsed since its last event:

$$\psi_i^{W}(\tilde{\tau}_i|\tilde{t}_i) = \frac{(\alpha - 1)(1 + \tilde{t}_i)^{\alpha-1}}{(1 + \tilde{t}_i + \tilde{\tau}_i)^{\alpha}}. \tag{5.4}$$

Due to the highly skewed shape of ψ_i, the expected waiting time until the next event becomes longer if more time has already elapsed without an event; that is, $\tilde{\tau}_i$ tends to be longer than τ_i. One can show this counterintuitive result by comparing the mean values of $\langle\tilde{\tau}_i\rangle$ and $\langle\tau_i\rangle$. The former is equal to $\langle\tilde{\tau}_i\rangle = \int_0^\infty \tau'\psi_i^{\mathrm{w}}(\tau'|\tilde{t}_i)d\tau' = (1 + \tilde{t}_i)/(\alpha - 2)$. This is larger than the latter, which is given by $\langle\tau_i\rangle = \int_0^\infty \tau'\psi_i(\tau')d\tau' = 1/(\alpha - 2)$.

When i is a Poisson process, ψ_i is an exponential distribution, and such a complication does not occur. The memoryless property of the exponential distribution yields $\psi_i^{\mathrm{w}}(\tilde{\tau}_i|\tilde{t}_i) = \psi_i(\tilde{\tau}_i)$, which we can verify by plugging the exponential distribution into Eq. (5.1):

$$
\psi_i^{\mathrm{w}}(\tilde{\tau}_i|\tilde{t}_i) = \frac{\lambda e^{-\lambda(\tilde{t}_i+\tilde{\tau}_i)}}{e^{-\lambda\tilde{t}_i}}
$$

$$
= \lambda e^{-\lambda\tilde{\tau}_i}. \tag{5.5}
$$

Therefore, $\psi_i^{\mathrm{w}}(\tilde{\tau}_i|\tilde{t}_i)$ does not depend on the time elapsed since the last event, $\tilde{t}_i$, and is the same as the original exponential distribution, $\psi_i(\tilde{\tau}_i)$. The original Gillespie algorithm fully exploits this property of Poisson processes.

To build a direct Gillespie method for simulating coupled renewal processes, we need to calculate two quantities: (i) the time until the next event in the entire population, τ, whichever process produces this event; (ii) the probability Π_i that the next event is produced by the ith process. We denote by $\phi(\tau, i|\{\tilde{t}_j\})$ the probability density for the ith process, and not any other process, to generate the next event after a time τ conditioned on the time elapsed since the last event of all processes in the population, $\{\tilde{t}_j\} \equiv \{\tilde{t}_1, \ldots, \tilde{t}_M\}$. It should be noted that we need to condition on each $\tilde{t}_j$. This is because $\phi(\tau, i|\{\tilde{t}_j\})$ depends not only on the ith renewal process generating an event after the waiting time τ but also on all the other processes not generating any event during this time. By putting all this together, we obtain

$$
\phi(\tau, i|\{\tilde{t}_j\}) = \psi_i^{\mathrm{w}}(\tau|\tilde{t}_i) \prod_{j=1;j\neq i}^{M} \Psi_j(\tau|\tilde{t}_j), \tag{5.6}
$$

where $\Psi_j(\tau|\tilde{t}_j)$ is the conditional survival probability for the waiting time of the jth process if it were running in isolation, given that its last event occurred a time $\tilde{t}_j$ ago.

Equation (5.6) is composed of two factors. The first factor is the probability density for the ith process to generate the next event within a small time window around τ (i.e., between τ and $\tau + d\tau$ from now, where $d\tau$ is infinitesimally small), corresponding to the probability density $\psi_i^{\mathrm{w}}(\tau|\tilde{t}_i)$. The other factor is the probability that none of the other $M - 1$ processes generates an event within this time window, corresponding to the product of the survival probabilities $\Psi_j(\tau|\tilde{t}_j)$ over all $j \neq i$. Using Eq. (5.1), we obtain $\Psi_j(\tau|\tilde{t}_j)$ as follows:

$$\Psi_j(\tau|\tilde{t}_j) = \int_\tau^\infty \psi_j^W(\tau'|\tilde{t}_j) \, d\tau' = \frac{\Psi_j(\tilde{t}_j + \tau)}{\Psi_j(\tilde{t}_j)}. \tag{5.7}$$

By substituting Eqs. (5.1) and (5.7) into Eq. (5.6), we obtain

$$\phi(\tau, i|\{\tilde{t}_j\}) = \frac{\psi_i(\tilde{t}_i + \tau)}{\Psi_i(\tilde{t}_i + \tau)} \Phi(\tau|\{\tilde{t}_j\}), \tag{5.8}$$

where

$$\Phi(\tau|\{\tilde{t}_j\}) = \prod_{j=1}^M \frac{\Psi_j(\tilde{t}_j + \tau)}{\Psi_j(\tilde{t}_j)}. \tag{5.9}$$

We interpret Eq. (5.8) as follows.

First, $\Psi_j(\tilde{t}_j)$ is the probability that the jth renewal process has not gener-ated any event for a time $\tilde{t}_j$ since its last event. The factor $\Psi_j(\tilde{t}_j + \tau)$ is the probability that the same process has not generated any event for time $\tilde{t}_j$ since its last event and it does not generate any event for another time τ. Therefore, $\Psi_j(\tilde{t}_j + \tau)/\Psi_j(\tilde{t}_j)$ is the conditional probability that the jth process does not gen-erate any event during the next time τ given that a time $\tilde{t}_j$ has already elapsed since it generated its last event. Equation (5.9) gives the probability that none of the M processes produces an event for time τ. So, it is the survival probabil-ity for the entire population. In other words, it is the probability that the next event in the entire population occurs sometime after time τ from now.

Second, the factor $\psi_i(\tilde{t}_i + \tau)/\Psi_i(\tilde{t}_i + \tau)$ on the right-hand side of Eq. (5.8) is the probability density function that the ith process generates an event at a time $\tilde{t}_i + \tau$ since its last event given that it has not generated any event before this time since the last event. Only this factor creates the dependence of $\phi(\tau, i|\{\tilde{t}_j\})$ on i. Given this observation, we define

$$\Pi_i \equiv \frac{\phi(\tau, i|\{\tilde{t}_j\})}{\sum_{j=1}^M \phi(\tau, j|\{\tilde{t}_j\})} = \frac{\lambda_i(\tilde{t}_i + \tau)}{\sum_{j=1}^M \lambda_j(\tilde{t}_j + \tau)}, \tag{5.10}$$

where

$$\lambda_i(t) = \frac{\psi_i(t)}{\Psi_i(t)} \tag{5.11}$$

is the instantaneous rate of the ith process.

In the original Gillespie algorithm, we equated the survival probability of the next event time for the entire population to u, a random variate obeying a uniform density on $(0, 1]$, to produce τ using inverse sampling. Similarly, a non-Markovian Gillespie algorithm can use inverse sampling to produce τ based on Eq. (5.9). However, once a uniform random variate u is drawn, solving $\Phi(\tau|\{\tilde{t}_j\}) = u$ is time-consuming because one cannot explicitly solve $\Phi(\tau|\{\tilde{t}_j\}) = u$ for τ in general, and thus one must solve it by numerical integration to produce each single event. This restriction does not prevent the algorithm from working but makes it too slow to be of practical use in many cases.

The non-Markovian Gillespie algorithm resolves this issue as follows. We first rewrite Eq. (5.9) as

$$\Phi(\tau|\{\tilde{t}_j\}) = \exp\left[-\sum_{j=1}^{M} \ln \frac{\Psi_j(\tilde{t}_j)}{\Psi_j(\tilde{t}_j + \tau)}\right]. \tag{5.12}$$

When M is large, it is unlikely that no process generates an event during a long time interval. Therefore, the τ values realized as the solution of $\Phi(\tau|\{\tilde{t}_j\}) = u$ will generally be small. This is equivalent to the situation in which $\Phi(\tau|\{\tilde{t}_j\})$ is tiny except for $\tau \approx 0$. Based on this observation, we approximate Eq. (5.12) by a first-order cumulant expansion around $\tau = 0$. This is done by the substitution of the following Taylor expansion of $\Psi_j(\tilde{t}_j + \tau)$:

$$\Psi_j(\tilde{t}_j + \tau) = \Psi_j(\tilde{t}_j) - \psi_j(\tilde{t}_j)\tau + O(\tau^2), \tag{5.13}$$

for $j = 1, \ldots, M$, into Eq. (5.12). This substitution leads to the following simplified expression for $\Phi(\tau|\{\tilde{t}_j\})$:

$$\begin{aligned}\Phi(\tau|\{\tilde{t}_j\}) &= \exp\left[-\sum_{j=1}^{M} \ln \frac{\Psi_j(\tilde{t}_j)}{\Psi_j(\tilde{t}_j) - \psi_j(\tilde{t}_j)\tau + O(\tau^2)}\right] \\ &= \exp\left\{-\sum_{j=1}^{M} \ln\left[1 + \frac{\psi_j(\tilde{t}_j)}{\Psi_j(\tilde{t}_j)}\tau + O(\tau^2)\right]\right\} \\ &= \exp\left[-\sum_{j=1}^{M} \frac{\psi_j(\tilde{t}_j)}{\Psi_j(\tilde{t}_j)}\tau + O(\tau^2)\right] \\ &\approx \exp\left[-\tau M \bar{\lambda}(\{\tilde{t}_j\})\right], \end{aligned} \tag{5.14}$$

where

$$\bar{\lambda}(\{\tilde{t}_j\}) = \frac{\sum_{j=1}^{M} \lambda_j(\tilde{t}_j)}{M} = \frac{1}{M}\sum_{j=1}^{M} \frac{\psi_j(\tilde{t}_j)}{\Psi_j(\tilde{t}_j)}. \tag{5.15}$$

The variable $\bar{\lambda}(\{\tilde{t}_j\})$ is the average instantaneous event rate. By instantaneous, we mean that the event rate changes over time even if no event has happened, which contrasts with the situation of the Poisson processes. The variant of a Poisson process in which the event rate varies over time is called the nonhomogeneous Poisson process. However, the non-Markovian Gillespie algorithm assumes that the event rate $\bar{\lambda}(\{\tilde{t}_j\})$ stays constant until the next event occurs somewhere in the coupled renewal processes. This is justified because the time to the next event, τ, is small when M is large, and therefore the change in $\bar{\lambda}(\{\tilde{t}_j\})$ should be negligible. See Legault and Melbourne (2019) for an application of the same idea to stochastic population dynamics in ecology when the environment is dynamically changing.

Note that the Taylor expansion given by Eq. (5.13) assumes that all $\Psi_j(\tilde{t}_j)$ are analytical at $\tilde{t}_j = 0$. This is not always the case in practice, which may cause some terms to diverge in the Taylor expansion of $\Psi_i(\tilde{t}_i + \tau)$, where i is the process that has generated the last event. To deal with this, the authors proposed to simply remove the renewal process that has generated the last event from the summation in Eq. (5.15).

We determine the time to the next event by solving $\Phi(\tau|\{\tilde{t}_j\}) = u$ for τ using the approximation given by Eq. (5.14). By doing this, we obtain

$$\tau = -\frac{\ln u}{M\bar{\lambda}(\{\tilde{t}_j\})}. \tag{5.16}$$

Now, the computation of τ is as fast as that for the original Gillespie algorithm except that the computation of $\bar{\lambda}(\{\tilde{t}_j\})$ may be complicated to some extent. Because τ should be small when M is large, one determines the process that generates this event by setting $\tau = 0$ in Eq. (5.10), that is,

$$\Pi_i = \frac{\lambda_i(\tilde{t}_i)}{M\bar{\lambda}(\{\tilde{t}_j\})}. \tag{5.17}$$

Equations (5.16) and (5.17) define the non-Markovian Gillespie algorithm (Boguñá et al., 2014). For Poisson processes, we have $\lambda_i(\tilde{t}_i) = \lambda_i$, and we recover the original direct method, which is given by Eqs. (3.8) and (3.9). Because the non-Markovian Gillespie algorithm assumes large M, its accuracy is considered to be good for large M.

By putting together these results we can define an extension of the direct method of Gillespie to simulate coupled renewal processes. The algorithm is described in Box 8. For simplicity, we have assumed so-called ordinary renewal processes, in which all processes have had the last event at $t = 0$ (Cox, 1962; Masuda & Lambiotte, 2020).

In the context of chemical reaction systems, Carletti and Filisetti (2012) developed a second-order variant of the Gillespie algorithm for chemical reactions in a dynamically varying volume, which one can also use to simulate more general non-Markovian processes. The non-Markovian Gillespie algorithm we presented in this section can be seen as a first-order algorithm in terms of τ (see Eqs. (5.13) and (5.14)). The idea of the second-order variant is to use the

---

Box 8 Non-Markovian Gillespie Algorithm

0. Initialization:

   (a) Define the initial state of the system, and set $t = 0$.

   (b) Initialize $\tilde{t}_j = 0$ for all $j \in \{1, \ldots, M\}$.

   (c) Calculate the rate $\lambda_j(\tilde{t}_j)$ for all j.

   (d) Calculate $\overline{\lambda}(\{\tilde{t}_j\}) = \sum_{j=1}^{M} \lambda_j(\tilde{t}_j)/M$.

1. Draw a uniform random variate u_1 from $(0, 1]$, and generate the waiting time to the next event by $\tau = -\ln u_1 / [M\overline{\lambda}(\{\tilde{t}_j\})]$.

2. Draw u_2 from a uniform distribution on $(0, M\overline{\lambda}(\{\tilde{t}_j\})]$. Select the event i to occur by iterating over $i = 1, \ldots, M$ until we find the i for which $\sum_{j=1}^{i-1} \lambda_j(\tilde{t}_j) < u_2 \leq \sum_{j=1}^{i} \lambda_j(\tilde{t}_j)$.

3. Perform the event on reaction channel i.

4. Advance the time according to $t \rightarrow t + \tau$.

5. (a) Update the list of times since the last event as $\tilde{t}_j \rightarrow \tilde{t}_j + \tau$ for all $j \neq i$, and set $\tilde{t}_i = 0$.

   (b) If there are processes j whose distribution of interevent times has changed upon the event, update $\psi_j(\tau)$ for the affected j values, and reset $\tilde{t}_j = 0$ if necessary.

   (c) Update $\lambda_j(\tilde{t}_j)$ for all $j \in \{1, \ldots, M\}$ as well as $\overline{\lambda}(\{\tilde{t}_j\}) = \sum_{j=1}^{M} \lambda_j(\tilde{t}_j)/M$.

6. Return to Step 1.

Taylor expansion of $\Psi_j(\tilde{t}_j + \tau)$ up to the second order, namely

$$\Psi_j(\tilde{t}_j + \tau) = \Psi_j(\tilde{t}_j) - \psi_j(\tilde{t}_j)\tau - \frac{\psi_j'(\tilde{t}_j)\tau^2}{2} + O(\tau^3), \tag{5.18}$$

instead of Eq. (5.13). Then, one obtains

$$u = \Phi(\tau | \{\tilde{t}_j\}) \approx \exp\left[-\tau M\overline{\lambda}(\{\tilde{t}_j\}) + c\tau^2\right], \tag{5.19}$$

where c is a constant. We refer to Carletti and Filisetti (2012) for its precise form. Therefore, we set τ by solving the quadratic equation in terms of τ:

$$c\tau^2 - M\overline{\lambda}(\{\tilde{t}_j\})\tau - \ln u = 0. \tag{5.20}$$

This second-order approximation should generally be a more accurate approximation than the first-order one. However, it comes at an increased computational cost for calculating c. Furthermore, it still relies on an assumption of large M, which breaks down when M is of the order of 1. For contagion processes, for example, M is typically of the order of 1 at the start or near the end of a simulation where it is often the case that only one or a few individuals are infectious.

5.3 Laplace Gillespie Algorithm

In spite of the approximations made by the non-Markovian Gillespie algorithm to make it fast enough for practical applications, the algorithm still requires that we update the instantaneous event rates of all processes whenever an event occurs (Step 5(c) in Box 8). This makes its runtime linear in terms of the number of reaction channels M. Note that we cannot lessen the time complexity by using any of the advanced methods discussed in Section 4. This is because these advanced methods only improve the efficiency of selecting the reaction channel in Step 2 in Box 8 and leave the overall time complexity of the entire algorithm to be linear. Note that using the binary search tree (Section 4.3) makes the algorithm less efficient because all nodes in the binary tree then need to be updated after each event, resulting in an algorithm with $O(M \log M)$ time complexity.

In this section, we explain an alternative generalization of the direct method to simulate non-Poissonian renewal processes. The algorithm, which we call the *Laplace Gillespie algorithm*, exploits mathematical properties of the Laplace transform (Masuda & Rocha, 2018) to allow exact and fast simulation of non-Poissonian renewal processes with fat-tailed waiting-time distributions. It takes advantage of the fact that a fat-tailed distribution often can be expressed as a mixture of exponential distributions. In other words, an appropriately weighted average of $\lambda e^{-\lambda \tau}$ over different values of λ approximates a desired fat-tailed distribution well. This situation is schematically shown in Fig. 23. As nothing ever comes for free, the Laplace Gillespie algorithm does not work for simulating arbitrary renewal processes. It only works for renewal processes whose distribution of interevent times satisfies a condition known as *complete monotocity*, which we discuss in detail at the end of this section. Luckily, fat-tailed distributions of interevent times that are ubiquitous in human interaction dynamics are often well modeled by completely monotone functions, so that the Laplace Gillespie algorithm is broadly applicable to social systems.

To explain the Laplace Gillespie algorithm, we first consider a single renewal process, which has an associated probability density function of interevent times $\psi(\tau)$. Our aim is to (repeatedly) produce interevent times, τ, that obey the probability density $\psi(\tau)$. To this end, we first draw a rate of a Poisson process, denoted by λ, from a fixed probability density $p(\lambda)$. Second, we draw the next value of τ from the exponential distribution $\lambda e^{-\lambda \tau}$ as if we were running a Poisson process with rate λ. Third, we advance the clock by τ and produce the

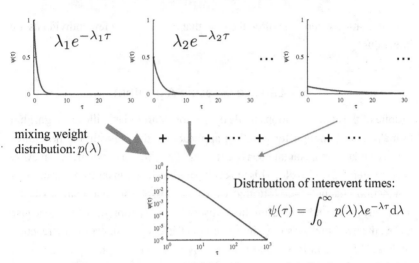

Figure 23 Schematic showing a mixture of exponential distributions and the mechanism of the Laplace Gillespie algorithm. One draws λ_1, λ_2, and so forth from $p(\lambda)$. The probability density function $p(\lambda)$ is called the mixing weight distribution, representing how probable each value of λ is to be drawn. Once a λ value is drawn, one generates the time to the next time, τ, according to the exponential distribution $\lambda e^{-\lambda \tau}$. As a result, one mixes exponential distributions with mixing weights $p(\lambda)$ to obtain the distribution of interevent times, $\psi(\tau)$. Although each component distribution is an exponential distribution, the mixture may yield a fat-tailed distribution.

event. Fourth, we repeat the procedure to determine the time to the next event. In other words, we redraw the rate, which we denote by λ' to avoid confusion, from the probability density $p(\lambda)$ and generate the time to the next event from the exponential distribution $\lambda' e^{-\lambda' \tau}$.

If the λ value drawn from $p(\lambda)$ happens to be large, then, τ tends to be small, and vice versa. Because there is diversity in the value of λ, the eventual distribution of interevent times, $\psi(\tau)$, is more dispersed than a single exponential distribution (Yannaros, 1994).

For a given $p(\lambda)$, the process generated by this algorithm is a renewal process. By construction, $\psi(\tau)$ is the mixture of exponential distributions given by

$$\psi(\tau) = \int_0^\infty p(\lambda)\lambda e^{-\lambda \tau}\, d\lambda. \tag{5.21}$$

For example, if there are only two possible values of λ, namely λ_{low} and λ_{high} ($> \lambda_{\text{low}}$), one obtains

$$p(\lambda) = q\delta(\lambda - \lambda_{\text{low}}) + (1 - q)\delta(\lambda - \lambda_{\text{high}}), \tag{5.22}$$

where δ is the Dirac delta function. Equation (5.22) just says that $\lambda = \lambda_{\text{low}}$ occurs with probability q and $\lambda = \lambda_{\text{high}}$ occurs with probability $1 - q$. Inserting Eq. (5.22) in Eq. (5.21) yields

$$\psi(\tau) = q\lambda_{\text{low}}e^{-\lambda_{\text{low}}\tau} + (1 - q)\lambda_{\text{high}}e^{-\lambda_{\text{high}}\tau}, \tag{5.23}$$

namely a mixture of two exponential distributions. (See Fonseca dos Reis et al. [2020]; Jiang et al. [2016]; Masuda and Holme [2020]; and Okada, Yamanishi, and Masuda [2020] for analysis of interevent times with a mixture of two exponential distributions.)

As another example, let us consider the gamma distribution for the distribution of mixing weights, that is,

$$p(\lambda) = \frac{\lambda^{\alpha-1}e^{-\lambda/\kappa}}{\Gamma(\alpha)\kappa^{\alpha}}, \tag{5.24}$$

where α and κ are the shape and scale parameters of the gamma distribution, respectively, and

$$\Gamma(\alpha) = \int_{0}^{\infty} x^{\alpha-1}e^{-x}dx \tag{5.25}$$

is the gamma function. Inserting Eq. (5.24) in Eq. (5.21) yields

$$\psi(\tau) = \frac{\kappa\alpha}{(1 + \kappa\tau)^{\alpha+1}}, \tag{5.26}$$

which is a power-law distribution (see the solid line in Fig. 24 for an example). Crucially, this example shows that one can create a power-law distribution, which is fat-tailed, by appropriately mixing exponential distributions, which are not fat-tailed.

What we want to really simulate is an ensemble of M simultaneously ongoing renewal processes that are governed by given distributions of interevent times, $\psi_i(\tau)$ $(i = 1,\ldots,M)$. We suppose that we can realize each $\psi_i(\tau)$ as a mixture of exponential distributions by appropriately setting a distribution of mixing weights $p_i(\lambda)$. In other words, we assume that we can find $p_i(\lambda)$ satisfying $\psi_i(\tau) = \int_{0}^{\infty} p_i(\lambda)\lambda e^{-\lambda\tau}d\lambda$. The Laplace Gillespie algorithm for simulating such a system is described in Box 9.

In contrast to the non-Markovian Gillespie algorithm, the Laplace Gillespie algorithm is exact for arbitrary values of M. Figure 25 shows an example in which the Laplace Gillespie algorithm is considerably more accurate than the non-Markovian Gillespie algorithm when $M = 10$ (Fig. 25(a)), whereas both algorithms are sufficiently accurate when $M = 100$ (Fig. 25(b)). In addition, the Laplace Gillespie algorithm tends to be faster than the non-Markovian Gillespie algorithm (Masuda & Rocha, 2018) because it does not need to update all the λ_i values (with $i = 1,\ldots,M$) after each event. Used together with the binary tree

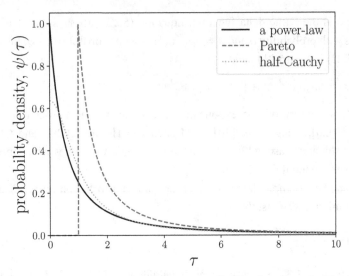

Figure 24 Three power-law distributions: the distribution given by Eq. (5.26) with $\alpha = 1$ and $\kappa = 1$, a Pareto distribution with $\alpha = 1$ and $\tau_0 = 1$, and a half-Cauchy distribution. Note that the three distributions follow the same asymptotic power law, $\psi(\tau) \propto \tau^{-2}$, as $\tau \to \infty$.

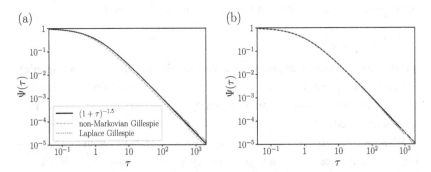

Figure 25 Comparison between the non-Markovian Gillespie algorithm and the Laplace Gillespie algorithm. We use a power-law distribution of interevent times $\psi(\tau) = \alpha/(1 + \tau)^{\alpha+1}$ with $\alpha = 1.5$. (a) $M = 10$. (b) $M = 100$. For each algorithm, the survival function of the interevent time distribution is plotted for just one of the M processes and compared against the ground truth, that is, $\Psi(\tau) = 1/(1 + \tau)^{\alpha}$.

structure (Section 4.3) or the composition and rejection method (Section 4.5), the Laplace Gillespie algorithm can thus simulate coupled renewal processes in $O(\log M)$ or $O(1)$ time.

Not all functional forms for $\psi_i(\lambda)$ can be generated as a mixture of exponentials. In these cases we cannot use the Laplace Gillespie algorithm. By contrast, one can use the non-Markovian Gillespie algorithm for any $\psi_i(\tau)$ in principle.

Box 9 Laplace Gillespie Algorithm

0. Initialization:

    (a) Define the initial state of the system, and set $t = 0$.

    (b) Initialize each of the M renewal processes by drawing the rate λ_j of the jth Poisson process from $p_j(\lambda_j)$ for all $j \in \{1, \ldots, M\}$.

    (c) Calculate the total event rate $\Lambda = \sum_{j=1}^{M} \lambda_j$.

(1) Draw a random variate u from a uniform distribution on $(0, 1]$, and generate the waiting time to the next event by $\tau = -\ln u / \Lambda$.

(2) Select the process that generates the next event with probability $\Pi_i = \lambda_i / \Lambda$.

(3) Implement the event taking place on the ith renewal process.

(4) Advance the clock according to $t \rightarrow t + \tau$.

(5) (a) Update $p_i(\lambda_i)$ if it has changed following the event.

    (b) Redraw a rate λ_i according to $p_i(\lambda_i)$.

    (c) If there are other processes j whose distribution of interevent times has changed following the event on the ith process, update each affected $p_j(\lambda_j)$ and redraw λ_j from the new $p_j(\lambda_j)$. The event rates of the other processes remain unchanged.

    (d) Update the total event rate $\Lambda = \sum_{j=1}^{M} \lambda_j$.

6. Return to Step (1).

To examine more formally to which cases the Laplace Gillespie algorithm is applicable, we integrate both sides of Eq. (5.21) to obtain

$$\Psi(\tau) = \int_{\tau}^{\infty} \psi(\tau') \mathrm{d}\tau' = \int_{0}^{\infty} p(\lambda) e^{-\lambda \tau} \mathrm{d}\lambda. \tag{5.27}$$

Equation (5.27) indicates that the survival probability of interevent times, $\Psi(\tau)$, is the Laplace transform of $p(\lambda)$. Therefore, the Laplace Gillespie algorithm can simulate a renewal process if and only if $\Psi(\tau)$ is the Laplace transform of a probability density function on non-negative values.

It is mathematically known that a necessary and sufficient condition for the existence of $p(\lambda)$ is that $\Psi(\tau)$ is completely monotone (Feller, 1971) and that $\Psi(0) = 1$. A function $\Psi(\tau)$ is said to be completely monotone if

$$(-1)^n \frac{\mathrm{d}^n \Psi(\tau)}{\mathrm{d}\tau^n} \geq 0 \quad (\tau \geq 0, n = 0, 1, \ldots). \tag{5.28}$$

The condition $\Psi(0) = \int_0^\infty \psi(\tau) \mathrm{d}\tau = 1$ is always satisfied since Ψ is a survival function. Equations (5.28) with $n = 0$ and $n = 1$ read $\Psi(\tau) \geq 0$ and $\psi(\tau) \geq 0$, respectively. These two inequalities are also always satisfied. Equation (5.28) offers nontrivial conditions when $n \geq 2$. For example, the condition with $n = 2$ reads

$$(-1)^2 \frac{d^2 \Psi(\tau)}{d\tau^2} = -\frac{d\psi(\tau)}{d\tau} \geq 0, \tag{5.29}$$

that is, $d\psi(\tau)/d\tau \leq 0$. Therefore, $\psi(\tau)$ must monotonically decrease with respect to τ. This condition excludes the Pareto distribution, which is a popular form of power-law distribution,

$$\psi(\tau) = \begin{cases} \frac{\alpha}{\tau_0} \left(\frac{\tau_0}{\tau}\right)^{\alpha+1} & (\tau \geq \tau_0), \\ 0 & (\tau < \tau_0), \end{cases} \tag{5.30}$$

where $\alpha > 0$ and $\tau_0 > 0$. We show the Pareto distribution with $\alpha = 1$ and $\tau_0 = 1$ by the dashed line in Fig. 24. Note that $\psi(\tau)$ discontinuously increases as τ increases across $\tau = \tau_0$; note that the Pareto distribution is defined for $\tau \geq 0$ (and $\psi(\tau) = 0$ for $0 \leq \tau < \tau_0$). Therefore, one cannot use the Laplace Gillespie algorithm when any $\psi_i(\tau)$ is a Pareto distribution.

To show another example of disqualified $\psi(\tau)$, consider Eq. (5.28) for $n = 3$, that is, $d\psi^2(\tau)/d\tau^2 \geq 0$. The half-Cauchy distribution, which is another form of power-law distribution, defined by

$$\psi(\tau) = \frac{2}{\pi(\tau^2 + 1)}, \tag{5.31}$$

where $\tau \geq 0$, violates this condition. (See the red dotted line in Fig. 24 for a plot.) This is because $d^2\psi(\tau)/d\tau^2 = 4(3\tau^2 - 1)/[\pi(\tau^2 + 1)^3]$, whose sign depends on the value of τ. Specifically, the half-Cauchy distribution has an inflection point at $\tau = 1/\sqrt{3}$. Note that the half-Cauchy distribution satisfies the condition for $n = 2$ (Eq. (5.29)) because $d\psi(\tau)/d\tau = -4\tau/[\pi(\tau^2 + 1)]$ < 0.

Complete monotonicity implies that the coefficient of variation (CV), defined by the standard deviation divided by the mean, of $\psi(\tau)$ is larger than or equal to 1 (Yannaros, 1994). This is natural because an exponential distribution, $\lambda e^{-\lambda\tau}$, has a CV equal to one. Because we are mixing exponential distributions with different λ values, the CV for the mixture of exponential distributions must be at least 1. This necessary condition for complete monotonicity excludes some distributions having less dispersion (i.e., standard deviation) than exponential distributions.

We stated various negative scenarios, but there are many distributions of interevent times, $\psi(\tau)$, for which the Laplace Gillespie algorithm works. The power-law distribution given by Eq. (5.26) is qualified because one can find the corresponding distribution of mixing weights, which is given by Eq. (5.24). In fact, using Eq. (5.26), we obtain

$$\Psi(\tau) = \int_\tau^\infty \psi(\tau')\mathrm{d}\tau' = \frac{1}{(1 + \kappa\tau)^\alpha}. \tag{5.32}$$

It is easy to verify that this $\Psi(\tau)$ is a completely monotone function.

As a second example, assume that $p(\lambda)$ is a uniform density on $[\lambda_{\min}, \lambda_{\max}]$ (Hidalgo, 2006). By Laplace transforming $p(\lambda)$ using Eq. (5.27), we obtain

$$\Psi(\tau) = \frac{e^{-\lambda_{\min}\tau} - e^{-\lambda_{\max}\tau}}{\tau(\lambda_{\max} - \lambda_{\min})}, \tag{5.33}$$

which yields

$$\psi(\tau) = -\frac{\mathrm{d}\Psi(\tau)}{\mathrm{d}\tau} = \frac{\lambda_{\min}e^{-\lambda_{\min}\tau} - \lambda_{\max}e^{-\lambda_{\max}\tau}}{(\lambda_{\max} - \lambda_{\min})\tau} + \frac{e^{-\lambda_{\min}\tau} - e^{-\lambda_{\max}\tau}}{(\lambda_{\max} - \lambda_{\min})\tau^2}. \tag{5.34}$$

Assume that $\lambda_{\min} \ll \lambda_{\max}$. If $\lambda_{\min} > 0$, then $\psi(\tau) \propto e^{-\lambda_{\min}\tau}/\tau$ as $\tau \to \infty$, which is a power-law distribution with an exponential cutoff, often resembling empirical data. If $\lambda_{\min} = 0$, then $\psi(\tau) \propto 1/\tau^2$ as $\tau \to \infty$.

A third example is when the interevent time obeys a gamma distribution:

$$\psi(\tau) = \frac{\tau^{\alpha-1}e^{-\tau/\kappa}}{\Gamma(\alpha)\kappa^\alpha}. \tag{5.35}$$

For this $\psi(\tau)$, we can express $\Psi(\tau)$ as the Laplace transform of $p(\lambda)$ if and only if $0 < \alpha \le 1$, and $p(\lambda)$ is given by

$$p(\lambda) = \begin{cases} 0 & (0 < \lambda < \kappa^{-1}), \\ \dfrac{1}{\Gamma(\alpha)\Gamma(1 - \alpha)\lambda(\kappa\lambda - 1)^\alpha} & (\lambda \ge \kappa^{-1}). \end{cases} \tag{5.36}$$

It is easy to verify that one obtains the exponential distribution by setting $\alpha = 1$ in Eq. (5.35). We refer to Masuda and Rocha (2018) for more examples of renewal processes that the Laplace Gillespie algorithm can simulate.

5.4 Temporal Gillespie Algorithm

The *temporal Gillespie algorithm* (Vestergaard & Génois, 2015) is an adaptation of the direct method to simulate coupled jump processes taking place on switching temporal networks, namely networks whose structure changes discontinuously in discrete points in time (Fig. 20). For simplicity in the following presentation, and without loss of generality, we furthermore assume that both the network's dynamics and the dynamical process start at time $t = 0$.

The starting point for developing a temporal Gillespie algorithm is a single isolated jump process i, which has a time-varying event rate $\lambda_i(t, \tilde{t}_i)$. Note that $\lambda_i(t, \tilde{t}_i)$ may depend on both the "wall clock" time, t, and the time since the last event, $\tilde{t}_i$. The explicit dependence on time t is not a property of the renewal processes considered in Sections 5.2 and 5.3, for which the event rate depends only on $\tilde{t}_i$. Conversely, if λ_i depends on t but not on $\tilde{t}_i$, the process is a nonhomogeneous Poisson process. In this Element we will only treat nonhomogeneous Poisson processes because they are simpler than the full problem in which the λ_i depends on both t and $\tilde{t}_i$. The derivation of the temporal Gillespie algorithm for the general case follows the same reasoning as for the algorithm for nonhomogeneous Poisson processes, but the mathematics is a bit more involved. We thus do not show the details here but refer interested readers to Vestergaard and Génois (2015).

Similar to how we calculated the waiting-time distribution for a Poisson process in Section 2.4, we first consider a discrete-time process and then take the continuous-time limit. More precisely, we want to know the probability that the ith process does not generate an event in a given time window $[t, t + \tau)$ (i.e., its survival probability), which we denote by $\Psi_i(\tau; t)$. We approximate $\Psi_i(\tau; t)$ by subdividing the interval into r small time-steps of size $\delta t = \tau / r$ as follows:

$$\Psi_i(\tau; t) \approx \prod_{r'=0}^{r-1} [1 - \lambda_i(t + r'\,\delta t)\delta t]. \tag{5.37}$$

Taking the limit $\delta t \to 0$, we find the following exact expression for the survival probability using the exponential identity (Appendix):

$$\Psi_i(\tau; t) = \exp\left(-\int_0^\tau \lambda_i(t + \tau')\,d\tau'\right)$$

$$= \exp\left(-\int_t^{t+\tau} \lambda_i(\tau')\,d\tau'\right). \tag{5.38}$$

Equation (5.38) does not reduce to a simple exponential, except in the special case of a constant λ_i. It does not even reduce to an analytical expression in general. For example, in the SIR process on a predefined switching temporal network, the infection rate for a given susceptible node i changes whenever an edge appears or disappears between i and an infectious node. Say, if a susceptible node i has two infectious neighbors and now gets connected to another infectious neighbor, the rate with which i gets infected changes from 2β to 3β, where β is the infection rate per contact. This means that one cannot in general solve Eq. (5.38) analytically even for a simple constant-rate SIR process in a temporal network. Nevertheless, owing to the conditional independence property of the jump processes (see Section 2.5), we can still find a formal expression for the waiting time for the superposition of a set of M processes. The survival function for a set of M processes is simply the product of the individual survival functions. Let t^{last} be the time of the last event amongst all

M processes. The survival function for the waiting time τ until the next event amongst all the processes is

$$
\begin{aligned}
\Psi(\tau; t^{\text{last}}) &= \prod_{i=1}^{M} \Psi_i(\tau; t^{\text{last}}) \\
&= \prod_{i=1}^{M} \exp\left(-\int_{t^{\text{last}}}^{t^{\text{last}}+\tau} \lambda_i(\tau')\,d\tau'\right) \\
&= \exp\left(-\int_{t^{\text{last}}}^{t^{\text{last}}+\tau} \sum_{i=1}^{M} \lambda_i(\tau')\,d\tau'\right) \\
&= \exp\left(-\int_{t^{\text{last}}}^{t^{\text{last}}+\tau} \Lambda(\tau')\,d\tau'\right),
\end{aligned}
\tag{5.39}
$$

where we have defined the *total instantaneous rate* as

$$
\Lambda(t) \equiv \sum_{i=1}^{M} \lambda_i(t).
\tag{5.40}
$$

Note that $\Lambda(t)$ is simply M times the average instantaneous rate, namely $\Lambda(t) = M\overline{\lambda}(t)$ (see Section 5.2).

Due to the lack of an analytic expression for $\Lambda(t)$, we need to numerically integrate Eq. (5.39) to evaluate it. However, inverting Eq. (5.39) to directly draw the waiting time is computationally too expensive. To overcome this, the temporal Gillespie algorithm works instead with unitless, normalized waiting times. Given the waiting time, τ, we define the normalized waiting time, denoted by $\overline{\tau}$, as

$$
\overline{\tau} = \int_{t^{\text{last}}}^{t^{\text{last}}+\tau} \Lambda(\tau')\,d\tau'.
\tag{5.41}
$$

The normalized waiting time follows an exponential distribution with an expected value of one. Therefore, it is easy to generate it using inverse sampling, that is, by $\overline{\tau} = -\ln u$, where u is a uniform random variate on $(0, 1]$. The (wall-clock) waiting time τ is found as the solution to Eq. (5.41) given the $\overline{\tau}$ value that we have generated.

In fact, all we have done for the moment is to exchange one implicit equation (Eq. (5.39)) for another (Eq. (5.41)). However, Eq. (5.41) is linear in Λ, which makes it easier to solve and approximate numerically. This is, in particular, the case because we assumed that the temporal network changes only in discrete points in time, as schematically shown in Fig. 20. We let $t_0^{\text{net}} = 0$, and we denote by $t_1^{\text{net}}, t_2^{\text{net}}, \ldots$ the subsequent time points at which the temporal network changes. Then, $[t_{n-1}^{\text{net}}, t_n^{\text{net}})$ is the nth interval between network changes. Since the temporal network only changes in discrete steps, $\Lambda(t)$ is piecewise constant. Therefore, one can solve Eq. (5.41) iteratively as follows. Suppose that we are given the time of the last event, t^{last}, and we want to find the time of the next event denoted by $t^{next} = t^{last} + \tau$. If there is no event yet, and we want to find the time of the first event, then we regard that $t^{last} = 0 = t_0^{net}$. We start from the time interval $[t_{n^*-1}^{net}, t_{n^*}^{net})$ between two successive switches of the network in which the last event took place, namely the interval that satisfies $t_{n^*-1}^{net} \leq t^{last} < t_{n^*}^{net}$. We then sequentially check for each time interval $[t^{last}, t_{n^*}^{net}), [t_{n^*}^{net}, t_{n^*+1}^{net}), [t_{n^*+1}^{net}, t_{n^*+2}^{net}), \ldots,$ to determine in which interval t^{next} falls. In practice, we compare at each step the generated value of τ to the value of the integral $\int_{t^{last}}^{t_n^{net}} \Lambda(t)\, dt$. The latter is efficiently calculated as the sum

$$\int_{t^{last}}^{t_n^{net}} \Lambda(t)\, dt = (t_{n^*}^{net} - t^{last})\Lambda_{n^*} + \sum_{\ell=n^*+1}^{n} \Delta_\ell \Lambda_\ell, \tag{5.42}$$

where $\Delta_\ell = t_\ell^{net} - t_{\ell-1}^{net}$ is the length of the ℓth interval between successive changes of the network, and Λ_ℓ is the value of $\Lambda(t)$ in this interval. If $n = n^*$, the sum in the second term on the right-hand side of Eq. (5.42) is the empty sum, that is, it has no summands and thus evaluates to zero, and the equation reduces to $\int_{t^{last}}^{t_{n^*}^{net}} \Lambda(t)\, dt = (t_{n^*}^{net} - t^{last})\Lambda_{n^*}$.

The smallest value of n that satisfies $\int_{t^{last}}^{t_n^{net}} \Lambda(t)\, dt > \tau$ determines the time interval in which the next event takes place. With that n value, the precise time of the next event is given by

$$t^{next} = t_{n-1} + \frac{\tau - \int_{t^{last}}^{t_{n-1}^{net}} \Lambda(t)\, dt}{\Lambda_n}. \tag{5.43}$$

Finally, we draw the Poisson process i that produces the event at time t^{next} with probability

$$\Pi_i(t^{next}) = \frac{\lambda_i(t^{next})}{\Lambda(t^{next})}. \tag{5.44}$$

The steps of the temporal Gillespie algorithm are described in Box 10. It works by iterating over the list of times at which the network changes. Within each interval between the network's switches, it compares the normalized waiting time, τ, to the total instantaneous rate integrated over the time interval, $\Lambda_n \Delta_n$ (see Step 2). If τ is larger than or equal to $\Lambda_n \Delta_n$, then nothing happens, and one subtracts $\Lambda_n \Delta_n$ from τ and advances to the next interval, $n + 1$ (see Step 2(a)). Alternatively, if τ is smaller than $\Lambda_n \Delta_n$, then an event occurs within the nth time window $[t_{n-1}^{net}, t_n^{net})$ (see Step 2(b)). Then, the algorithm determines the timing of the event and selects the reaction channel to produce the event using any of the appropriate selection methods discussed earlier (see

Box 10 Temporal Gillespie algorithm.

0. Initialization:

   (a) Define the initial state of the system, and set $t = 0$.

   (b) Set $n = 1$ and $\Delta = t_1^{net} - t_0^{net}$.

   (c) Initialize the rates λ_j for all $j \in \{1, \ldots, M\}$.

   (d) Calculate the total rate $\Lambda = \sum_{j=1}^{M} \lambda_j$.

1. Draw a normalized waiting time $\bar{\tau} = -\ln u$, where u is a uniform random variate on $(0, 1]$.

2. Compare $\Lambda\Delta$ to $\bar{\tau}$:

   (a) If $\Lambda\Delta \leq \bar{\tau}$, then no reaction takes place in the nth time window.

      i. Set $\bar{\tau} \to \bar{\tau} - \Lambda\Delta$.

      ii. Advance to the next time window by setting $t \to t_n^{net}$ and $\Delta \to t_{n+1}^{net} - t_n^{net}$; update $n \to n + 1$.

      iii. Update all λ_j affected by changes in the temporal network, and update Λ accordingly.

      iv. Return to Step 2.

   (b) If $\Lambda\Delta > \bar{\tau}$, then an event takes place at time $t^{next} = t + \bar{\tau}/\Lambda$.

      i. Select the reaction channel i that produces the event with probability $\Pi_i = \lambda_i/\Lambda$.

      ii. Update the time as $t \to t^{next}$. Also update the remaining length of the present time window as $\Delta \to \Delta - \bar{\tau}/\Lambda$.

      iii. Update the rates λ_j that are affected by the event, and update Λ accordingly.

      iv. Return to Step 1.

Sections 3.3, 4.3, and 4.5). It then updates the system, draws a new normalized waiting time, and repeats the procedure.

Vestergaard & Génois (2015) furthermore proposed to adapt the temporal Gillespie algorithm to simulate non-Markovian processes in temporal networks. To make the algorithm computationally efficient, they proposed two approximations to solve Eq. (5.41) by simply iterating over the times t_n^{net} at which the network changes, as we did for nonhomogeneous Poisson processes. These approximations avoid having to use numerical integration to solve the implicit equation, which would make the algorithm slow for large systems.

The first approximation is to regard the total instantaneous rate $\Lambda(t, \{\tilde{t}_j\})$, which in the non-Markovian case can depend on the times since the last events for all M processes, as constant during each interval $[t_{n-1}^{net}, t_n^{net})$ between the consecutive changes in the network. This approximation is accurate when the network changes much faster than the total rate $\Lambda(t, \{\tilde{t}_j\})$ does, namely when

$$\frac{\Lambda(t_{n+1}^{\text{net}}, \{\tilde{t}_j\}) - \Lambda(t_n^{\text{net}}, \{\tilde{t}_j\})}{\Lambda(t_n^{\text{net}}, \{\tilde{t}_j\})} \ll 1, \tag{5.45}$$

where $\Lambda(t_{n+1}^{\text{net}}, \{\tilde{t}_j\}) - \Lambda(t_n^{\text{net}}, \{\tilde{t}_j\})$ is the change of $\Lambda(t, \{\tilde{t}_j\})$ between two successive intervals. When simulating spreading processes in temporal networks, the network dynamics are often much faster than the spreading dynamics in practice. For example, the timescale of recordings of physical proximity networks is typically of the order of seconds to minutes while the infection and recovery of flu-like diseases occur in the order of hours to days (Vestergaard & Génois, 2015). With this first approximation, one can directly apply Eq. (5.42) and use the same procedure as for the Poissonian case.

The second approximation is to use a first-order cumulant expansion, similar to the non-Markovian Gillespie algorithm (Section 5.2), in addition to the first approximation. It amounts to assuming that each λ_i is constant as long as no event takes place and no change of the network that directly affects the λ_i value takes place. One thus avoids having to update all the λ_i values each time we go to the next time window (i.e., from $[t_{n-1}^{\text{net}}, t_n^{\text{net}})$ to $[t_n^{\text{net}}, t_{n+1}^{\text{net}}))$, and the algorithm runs much faster. To increase the accuracy of the algorithm when the number of reaction channels M is small and the cumulant expansion is not accurate (e.g., at the start or near the end of an SIR process where only a few nodes are typically infectious), they proposed a heuristic approach, in which one updates λ_i only if the time elapsed since the last update of λ_i exceeds a given threshold δ. They proposed to choose the value of δ as a given fraction of the expected waiting time of a single reaction channel. Therefore, when M is large, the waiting time between events will almost never exceed δ, and the algorithm will be similar to the non-Markovian Gillespie algorithm. When M is small, the algorithm updates the λ_i more frequently, making it more accurate at an added computational cost.

With the above approximations, the application of the temporal Gillespie algorithm to general non-Markovian processes only slightly changes the implementation from that for the nonhomogeneous Poisson processes described in Box 10. We refer interested readers to Vestergaard and Génois (2015) for details.

5.5 Event-Driven Simulation of the SIR Process

Holme proposed another efficient event-based algorithm, related to the first reaction method, when the time-stamped contact events are given as data (Holme, 2021). Although the timing of the event is no longer stochastic, the overall dynamics are still stochastic. This is because, in the SIR model for example, infection upon each contact event occurs with a certain probability and recovery occurs as a Poisson process with rate μ. The efficiency of the Holme's algorithm comes from multiple factors. Suppose that the ith node is infected and its neighboring node, j, is susceptible. First, the algorithm tactically avoids searching all the contact events between i and j when determining the event with which i successfully infects j. Second, it uses the binary heap to maintain a carefully limited set of times of the events with which infection may occur between pairs of nodes. The corresponding code for simulating the SIR model, implemented in C with a Python wrapper, is available on Github (Holme, 2021).

6 Conclusions

The aim of this article has been twofold: to provide a tutorial of the standard Gillespie algorithms and to review recent Gillespie algorithms that improve upon their computational efficiency and extend their scope. While our emphasis and examples lean toward social multiagent dynamics in populations and networks, the applicability of the Gillespie algorithms and their variants is extensive. We believe that the present article is useful for students and researchers in various fields, such as epidemiology, ecology, control theory, artificial life, complexity sciences, and so on.

In fact, many models of adaptive networks, where the network change is induced by the change of the status of, for example, nodes, have been mostly described by ODEs and assume that the interaction strength between pairs of nodes varies in response to changes in individuals' behavior (Gross & Blasius, 2008; Gross & Sayama, 2009; Wang et al., 2015). If such changes occur in an event-driven manner, Gillespie algorithms are readily applicable. How to deploy and develop Gillespie algorithms and their variants to adaptive network scenarios is a practical concern.

We briefly discussed simulations on empirical time-stamped contact event data (Sections 5.4 and 5.5). In this setting, it is the given data that determines the times and edges (i.e., node pairs) of the events, which is contrary to the assumption of the Gillespie algorithms that jump-process models generate events. Despite the increasing demand of simulations on the given time-stamped contact event data, this is still an underexplored area of research. Vestergaard and Génois (2015) and Holme (2021) showed that ideas and techniques from the Gillespie algorithms are useful for such simulations although the developed algorithms are distinct from the historical Gillespie algorithms. This is another interesting area of future research.

Appendix

Exponential Identity

In this appendix, we prove the identity

$$\lim_{x \to 0}(1 + x)^{1/x} = e. \tag{A.1}$$

Because e^x is continuous in x, we obtain

$$\lim_{x \to 0}(1 + x)^{1/x} = e^{\lim_{x \to 0} \ln(1+x)/x}. \tag{A.2}$$

Thus, we can prove Eq. (A.1) by showing that $\lim_{x \to 0} \ln(1 + x)/x = 1$. We do this using l'Hôpital's rule as follows:

$$\lim_{x \to 0} \frac{\ln(1 + x)}{x} = \frac{\lim_{x \to 0} \frac{1}{1+x}}{\lim_{x \to 0} 1} = 1. \tag{A.3}$$

References

Anderson, D. F. (2007). A modified next reaction method for simulating chemical systems with time dependent propensities and delays. *J. Chem. Phys.*, *127*, 214107.

Anderson, R. M., & May, R. M. (1991). *Infectious Diseases of Humans.* Oxford: Oxford University Press.

Andersson, H., & Britton, T. (2000). *Stochastic Epidemic Models and Their Statistical Analysis.* New York: Springer.

Barrat, A., Barthélemy, M., & Vespignani, A. (2008). *Dynamical Processes on Complex Networks.* Cambridge: Cambridge University Press.

Barrio, M., Burrage, K., Leier, A., & Tian, T. (2006). Oscillatory regulation of Hes1: Discrete stochastic delay modelling and simulation. *PLoS Comput. Biol.*, *2*, e117.

Bartlett, M. S. (1953). Stochastic processes or the statistics of change. *J. R. Statist. Soc. C, 2*, 44–64.

Black, A. J., & McKane, A. J. (2012). Stochastic formulation of ecological models and their applications. *Trends Ecol. Evol.*, *27*, 337–345.

Blue, J. L., Beichl, I., & Sullivan, F. (1995). Faster Monte Carlo simulations. *Phys. Rev. E, 51*, R867–R868.

Boguñá, M., Lafuerza, L. F., Toral, R., & Serrano, M. Á. (2014). Simulating non-Markovian stochastic processes. *Phys. Rev. E, 90*, 042108.

Bortz, A. B., Kalos, M. H., & Lebowitz, J. L. (1975). A new algorithm for Monte Carlo simulation of Ising spin systems. *J. Comput. Phys.*, *17*, 10–18.

Bratsun, D., Volfson, D., Tsimring, L. S., & Hasty, J. (2005). Delay-induced stochastic oscillations in gene regulation. *Proc. Natl. Acad. Sci. USA, 102*, 14593–14598.

Britton, T. (2010). Stochastic epidemic models: A survey. *Math. Biosci, 225*, 24–35.

Brown, R. G., Eddelbuettel, D., & Bauer, D. (2021). *Dieharder: A random number test suite, Version 3.31.1.* Accessed on November 24, 2021, from http://webhome.phy.duke.edu/~rgb/General/dieharder.php.

Cai, X. (2007). Exact stochastic simulation of coupled chemical reactions with delays. *J. Chem. Phys.*, *126*, 124108.

Campbell, J. Y., Lo, A. W., & MacKinlay, A. C. (1997). *The Econometrics of Financial Markets.* Princeton, NJ: Princeton University Press.

Carletti, T., & Filisetti, A. (2012). The stochastic evolution of a protocell: The Gillespie algorithm in a dynamically varying volume. *Comput. Math. Methods Med., 2012*, 423627.

Castellano, C., Fortunato, S., & Loreto, V. (2009). Statistical physics of social dynamics. *Rev. Mod. Phys., 81*, 591–646.

Chen, J., Edelkamp, S., Elmasry, A., & Katajainen, J. (2012). In-place heap construction with optimized comparisons, moves, and cache misses. *Lecture Notes on Computer Science, 7464*, 259–270.

Clementi, A. E. F., Macci, C., Monti, A., Pasquale, F., & Silvestri, R. (2008). Flooding time in edge-Markovian dynamic graphs. In *Proceedings of the 27th ACM SIGACT-SIGOPS Annual Symposium on Principles of Distributed Computing (PODC'08)* (pp. 213–222).

Codling, E. A., Plank, M. J., & Benhamou, S. (2008). Random walk models in biology. *J. R. Soc. Interface, 5*, 813–834.

Colizza, V., Barrat, A., Barthélemy, M., & Vespignani, A. (2006). The role of the airline transportation network in the prediction and predictability of global epidemics. *Proc. Natl. Acad. Sci. USA, 103*, 2015–2020.

Colizza, V., Pastor-Satorras, R., & Vespignani, A. (2007). Reaction-diffusion processes and metapopulation models in heterogeneous networks. *Nat. Phys., 3*, 276–282.

Cornforth, D., Green, D. G., & Newth, D. (2005). Ordered asynchronous processes in multi-agent systems. *Physica D: Nonlinear Phenom., 204*, 70–82.

Cota, W., & Ferreira, S. C. (2017). Optimized Gillespie algorithms for the simulation of Markovian epidemic processes on large and heterogeneous networks. *Comput. Phys. Commun., 219*, 303–312.

Cox, D. R. (1962). *Renewal Theory*. York, UK: Methuen & Co. Ltd.

Daley, D. J., & Gani, J. (1999). *Epidemic Modelling: An Introduction*. Cambridge: Cambridge University Press.

de Arruda, G. F., Rodrigues, F. A., & Moreno, Y. (2018). Fundamentals of spreading processes in single and multilayer complex networks. *Phys. Rep., 756*, 1–59.

Diekmann, O., & Heesterbeek, J. A. P. (2000). *Mathematical Epidemiology of Infectious Diseases*. Chichester, UK: John Wiley & Sons, Ltd.

Dobrinevski, A., & Frey, E. (2012). Extinction in neutrally stable stochastic Lotka-Volterra models. *Phys. Rev. E, 85*, 051903.

Doob, J. L. (1942). Topics in the theory of Markoff chains. *Trans. Am. Math. Soc., 52*, 37–64.

Doob, J. L. (1945). Markoff chains: Denumerable case. *Trans. Am. Math. Soc., 58*, 455–473.

Eugster, P. T., Guerraoui, R., Kermarrec, A.-M., & Massoulié, L. (2004). Epidemic information dissemination in distributed systems. *Computer, 37,* 60–67.

Farrington, C. P., Kanaan, M. N., & Gay, N. J. (2003). Branching process models for surveillance of infectious diseases controlled by mass vaccination. *Biostatistics, 4,* 279–295.

Feller, W. (1971). *An Introduction to Probability Theory and Its Applications, Volume II* (2nd ed.). New York: John Wiley & Sons.

Fennell, P. G., Melnik, S., & Gleeson, J. P. (2016). Limitations of discrete-time approaches to continuous-time contagion dynamics. *Phys. Rev. E, 94,* 052125.

Fonseca dos Reis, E., Li, A., & Masuda, N. (2020). Generative models of simultaneously heavy-tailed distributions of inter-event times on nodes and edges. *Phys. Rev. E, 102,* 052303.

Fosdick, B. K., Larremore, D. B., Nishimura, J., & Ugander, J. (2018). Configuring random graph models with fixed degree sequences. *SIAM Rev., 60,* 315–355.

Gabbiani, F., & Cox, S. J. (2010). *Mathematics for Neuroscientists.* Amsterdam: Academic Press.

Gibson, M. A., & Bruck, J. (2000). Efficient exact stochastic simulation of chemical systems with many species and many channels. *J. Phys. Chem. A, 104,* 1876–1889.

Gillespie, D. T. (1976). A general method for numerically simulating the stochastic time evolution of coupled chemical reactions. *J. Comput. Phys., 22,* 403–434.

Gillespie, D. T. (1977). Exact stochastic simulation of coupled chemical reactions. *J. Phys. Chem., 81,* 2340–2361.

Gillespie, D. T. (2001). Approximate accelerated stochastic simulation of chemicallyreacting systems. *J. Chem. Phys., 115,* 1716–1733.

Gleeson, J. P., Onaga, T., Fennell, P. et al. (2021). Branching process descriptions of information cascades on Twitter. *J. Comp. Netw.* DOI: https://doi.org/10.1093/comnet/cnab002.

Gokhale, C. S., Papkou, A., Traulsen, A., & Schulenburg, H. (2013). Lotka–Volterra dynamics kills the Red Queen: Population size fluctuations and associated stochasticity dramatically change host-parasite coevolution. *BMC Evol. Biol., 13,* 254.

Gómez, S., Gómez-Gardeñes, J., Moreno, Y., & Arenas, A. (2011). Nonperturbative heterogeneous mean-field approach to epidemic spreading in complex networks. *Phys. Rev. E, 84,* 036105.

Goutsias, J., & Jenkinson, G. (2013). Markovian dynamics on complex reaction networks. *Phys. Rep., 529*, 199–264.

Greil, F., & Drossel, B. (2005). Dynamics of critical Kauffman networks under asynchronous stochastic update. *Phys. Rev. Lett., 95*, 048701.

Gross, T., & Blasius, B. (2008). Adaptive coevolutionary networks: A review. *J. R. Soc. Interface, 5*, 259–271.

Gross, T., & Sayama, H. (Eds.). (2009). *Adaptive Networks*. Berlin: Springer.

Hanski, I. (1998). Metapopulation dynamics. *Nature, 396*, 41–49.

Hanson, F. B. (2007). *Applied Stochastic Processes and Control for Jump- Diffusions: Modeling, Analysis and Computation*. Philadelphia: Society for Industrial and Applied Mathematics.

Haramoto, H., Matsumoto, M., Nishimura, T., Panneton, F., & L'Ecuyer, P. (2008). Efficient jump ahead for $\mathbb{F}_2$-linear random number generators. *INFORMS J. Comput., 20*, 385–390.

Hidalgo R., C. A. (2006). Conditions for the emergence of scaling in the inter-event time of uncorrelated and seasonal systems. *Physica A: Stat. Mech. Appl., 369*, 877–883.

Hofbauer, J., & Sigmund, K. (1988). *The Theory of Evolution and Dynamical Systems*. Cambridge: Cambridge University Press.

Holley, R. A., & Liggett, T. M. (1975). Ergodic theorems for weakly interacting infinite systems and the voter model. *Ann. Prob., 3*, 643–663.

Holme, P. (2015). Modern temporal network theory: A colloquium. *Eur. Phys. J. B, 88*, 234.

Holme, P. (2021). Fast and principled simulations of the SIR model on temporal networks. *PLoS ONE, 16*, e0246961.

Holme, P., & Saramäki, J. (2012). Temporal networks. *Phys. Rep., 519*, 97–125.

Holme, P., & Saramäki, J. (2013). *Temporal Networks*. Berlin: Springer.

Holme, P., & Saramäki, J. (2019). *Temporal Network Theory*. Cham: Springer.

Huberman, B. A., & Glance, N. S. (1993). Evolutionary games and computer-simulations. *Proc. Natl. Acad. Sci. U.S.A., 90*, 7716–7718.

Hufnagel, L., Brockmann, D., & Geisel, T. (2004). Forecast and control of epidemics in a globalized world. *Proc. Natl. Acad. Sci. U.S.A., 101*, 15124–15129.

Isella, L., Romano, M., Barrat, A. et al. (2011). Close encounters in a pediatric ward: Measuring face-to-face proximity and mixing patterns with wearable sensors. *PLoS ONE, 6*, e17144.

Jagers, P (1975). *Branching Processes with Biological Applications*. London: John Wiley & Sons.

Jiang, Z.-Q., Xie, W.-J., Li, M.-X., Zhou, W.-X., & Sornette, D. (2016). Two-state Markov-chain Poisson nature of individual cellphone call statistics. *J. Stat. Mech., 2016*, 073210.

Jones, D. (2010). *Good Practice in (Pseudo) Random Number Generation for Bioinformatics Applications.* Accessed on May 21, 2021, from www.cs.ucl.ac.uk/staff/d.jones/GoodPracticeRNG.pdf.

Karsai, M., Kivelä, M., Pan, R. K. et al. (2011). Small but slow world: How network topology and burstiness slow down spreading. *Phys. Rev. E, 83*, 025102(R).

Kendall, D. G. (1950). An artificial realization of a simple "birth-and-death" process. *J. R. Statist. Soc. B, 12*, 116–119.

Kermack, W. O., & McKendrick, A. G. (1927). A contribution to the mathematical theory of epidemics. *Proc. R. Soc. Lond. A, 115*, 700–721.

Kierzek, A. M. (2002). STOCKS: STOChastic Kinetic Simulations of biochemical systems with Gillespie algorithm. *Bioinformatics, 18*, 470–481.

Kiss, I. Z., Berthouze, L., Taylor, T. J., & Simon, P. L. (2012). Modelling approaches for simple dynamic networks and applications to disease transmission models. *Proc. R. Soc. A, 468*, 1332–1355.

Kiss, I. Z., Miller, J. C., & Simon, P. L. (2017a). *Mathematics of Epidemics on Networks.* Cham: Springer.

Kiss, I. Z., Miller, J. C., & Simon, P. L. (2017b). *Mathematics of Epidemics on Networks, Associated Python Software.* Accessed on November 24, 2021, from https://springer-math.github.io/Mathematics-of-Epidemics-on-Networks/.

Kivelä, M., Pan, R. K., Kaski, K. et al. (2012). Multiscale analysis of spreading in a large communication network. *J. Stat. Mech., 2012*, P03005.

Knuth, D. E. (1976). Big omicron and big omega and big theta. *ACM SIGACT News, 8*, 18–24.

Krapivsky, P. L., Redner, S., & Ben-Naim, E. (2010). A *Kinetic View of Statistical Physics.* Cambridge: Cambridge University Press.

L'Ecuyer, P., & Simard, R. (2007). TestU01: A C library for empirical testing of random number generators. *ACM Trans. Math. Software, 33*, 1–40.

Legault, G., & Melbourne, B. A. (2019). Accounting for environmental change in continuous-time stochastic population models. *Theor. Ecol, 12*, 31–48.

Liggett, T. M. (1999). *Stochastic Interacting Systems: Contact, Voter and Exclusion Processes.* New York: Springer.

Liggett, T. M. (2010). *Continuous Time Markov Processes: An Introduction.* Providence, RI: American Mathematical Society.

Lu, T., Volfson, D., Tsimring, L., & Hasty, J. (2004). Cellular growth and division in the Gillespie algorithm. *Syst. Biol., 1*, 121–128.

Mantegna, R. N., & Stanley, H. E. (2000). *An Introduction to Econophysics.* Cambridge: Cambridge University Press.

Marchetti, L., Priami, C., & Thanh, V. H. (2017). *Simulation Algorithms for Computational Systems Biology.* Cham: Springer.

Masuda, N., & Holme, P. (2013). Predicting and controlling infectious disease epidemics using temporal networks. *F1000Prime Reports, 5,* 6.

Masuda, N., & Holme, P. (2020). Small inter-event times govern epidemic spreading on networks. *Phys. Rev. Research, 2,* 023163.

Masuda, N., & Lambiotte, R. (2020). *A Guide to Temporal Networks* (2nd ed.). Singapore: World Scientific.

Masuda, N., Porter, M. A., & Lambiotte, R. (2017). Random walks and diffusion on networks. *Phys. Rep., 716–717,* 1–58.

Masuda, N., & Rocha, L. E. C. (2018). A Gillespie algorithm for non-Markovian stochastic processes. *SIAM Rev., 60,* 95–115.

Matsumoto, M. (2021). *Mersenne Twister with improved initialization.* Accessed on May 21, 2021, from www.math.sci.hiroshima-u.ac.jp/m-mat/MT/MT2002/emt19937ar.html.

Matsumoto, M., & Nishimura, T. (1998). Mersenne Twister: A 623- dimensionally equidistributed uniform pseudo-random number generator. *ACM Trans. Model. Comput. Simul., 8,* 3–30.

McGill, B. J., Etienne, R. S., Gray, J. S. et al. (2007). Species abundance distributions: Moving beyond single prediction theories to integration within an ecological framework. *Ecol. Lett., 10,* 995–1015.

Miritello, G., Moro, E., & Lara, R. (2011). Dynamical strength of social ties in information spreading. *Phys. Rev. E, 83,* 045102(R).

Mollison, D., Isham, V., & Grenfell, B. (1994). Epidemics: Models and data. *J. R. Statist. Soc. A, 157,* 115–149.

Murray, J. D. (2002). *Mathematical Biology I. An Introduction* (3rd ed.). New York: Springer.

Ogura, M., & Preciado, V. M. (2016). Stability of spreading processes over time-varying large-scale networks. *IEEE Trans. Netw. Sci. Eng., 3,* 44–57.

Okada, M., Yamanishi, K., & Masuda, N. (2020). Long-tailed distributions of inter-event times as mixtures of exponential distributions. *R. Soc. Open. Sci., 7,* 191643.

Okubo, A., & Levin, S. A. (2001). *Diffusion and Ecological Problems: Modern Perspectives* (2nd ed.). New York: Springer.

O'Neill, M. E. (2014). *PCG: A family of simple fast space-efficient statistically good algorithms for random number generation.* Accessed on November 24, 2021, from https://www.pcg-random.org/pdf/toms-oneill-pcg-family-v1.02.pdf.

Acknowledgments

We thank the SocioPatterns collaboration (see www.sociopatterns.org) for providing the experimental data set. N. M. thanks the following institutions for their financial support: the AFOSR European Office (under Grant No. FA9550-19-1-7024), the National Science Foundation (under Grant No. DMS-2052720), the Nakatani Foundation, the Sumitomo Foundation, and the Japan Science and Technology Agency (under Grant No. JPMJMS2021). C. L. V. was supported in part by the French government under management of Agence Nationale de la Recherche as part of the "Investissements d'avenir" program, reference ANR-19-P3IA-0001 (PRAIRIE 3IA Institute).

Cambridge Elements ☰

The Structure and Dynamics of Complex Networks

Guido Caldarelli
Ca' Foscari University of Venice

Guido Caldarelli is Full Professor of Theoretical Physics at Ca' Foscari University of Venice. Guido Caldarelli received his Ph.D. from SISSA, after which he held postdoctoral positions in the Department of Physics and School of Biology, University of Manchester, and the Theory of Condensed Matter Group, University of Cambridge. He also spent some time at the University of Fribourg in Switzerland, at École Normale Supérieure in Paris, and at the University of Barcelona. His main scientific activity (interest?) is the study of networks, mostly analysis and modelling, with applications from financial networks to social systems as in the case of disinformation. He is the author of more than 200 journal publications on the subject, and three books, and is the current President of the Complex Systems Society (2018 to 2021).

About the Series

This cutting-edge series provides authoritative and detailed coverage of the underlying theory of complex networks, specifically their structure and dynamical properties. Each Element within the series will focus upon one of three primary topics: static networks, dynamical networks, and numerical/computing network resources.

Cambridge Elements $\equiv$

The Structure and Dynamics of Complex Networks

Printed in the United States
by Baker & Taylor Publisher Services